1/23

PLANTS
THAT
Cure

PLANTS
THAT
Cure

Plants as a Source for Medicines,
from Pharmaceuticals to
Herbal Remedies

Elizabeth A. Dauncey
and
Melanie-Jayne R. Howes

PRINCETON UNIVERSITY PRESS

PRINCETON AND OXFORD

Published in the United States and Canada in 2020 by Princeton University Press,
41 William Street, Princeton, New Jersey 08540
press.princeton.edu

Conceived, Designed and Produced by The Bright Press,
an imprint of The Quarto Group
The Old Brewery, 6 Blundell Street,
London N7 9BH, United Kingdom
T (0) 20 7700 6700 F (0)20 7700 8066
www.QuartoKnows.com

ISBN: 978-0-691-20018-7

Library of Congress Control Number: 2019950827

This book has been composed in Gentium and Frutiger
Printed on acid-free paper

Written by Dr Elizabeth A. Dauncey and Dr Melanie-Jayne R. Howes
Designed by Tony Seddon
Project managed by D & N Publishing, Wiltshire
Commissioned by Jacqui Sayers

Photograph page 2: opium poppy (*Papaver somniferum*)
Photograph opposite: common snowdrop (*Galanthus nivalis*)

Printed in Singapore

10 9 8 7 6 5 4 3 2 1

Disclaimer
The information in this book is intended to educate, delight and expand the reader's understanding of the diversity of plant life, the compounds plants produce and the uses to which humans have put these compounds as medicines. It does not purport to be, nor is it intended to be, a medical manual or a self-treatment guide for the use of medicinal or other plants. The authors and publishers make no claims about the efficacy of the plants and compounds mentioned in the book or whether they are able to cure or alleviate the illness or conditions that are discussed. The authors and publishers do not endorse the use of these plants for any of the applications described and are not responsible for any consequences arising from the use of this information for whatever reason, including curiosity or malicious or illegal intent. In matters of your health care, we recommend that you consult a qualified medical professional. If you suspect that you have a medical problem, we urge you to seek competent medical help.

Contents

INTRODUCTION   8

CHAPTER 1
THE BOTANICAL MEDICINE CHEST   10

CHAPTER 2
STRONG AT HEART   34

CHAPTER 3
CALMING THE NERVES   52

CHAPTER 4
IN SICKNESS AND IN HEALTH   78

CHAPTER 5
SUPPORTING THE ORGANS AND GLANDS   100

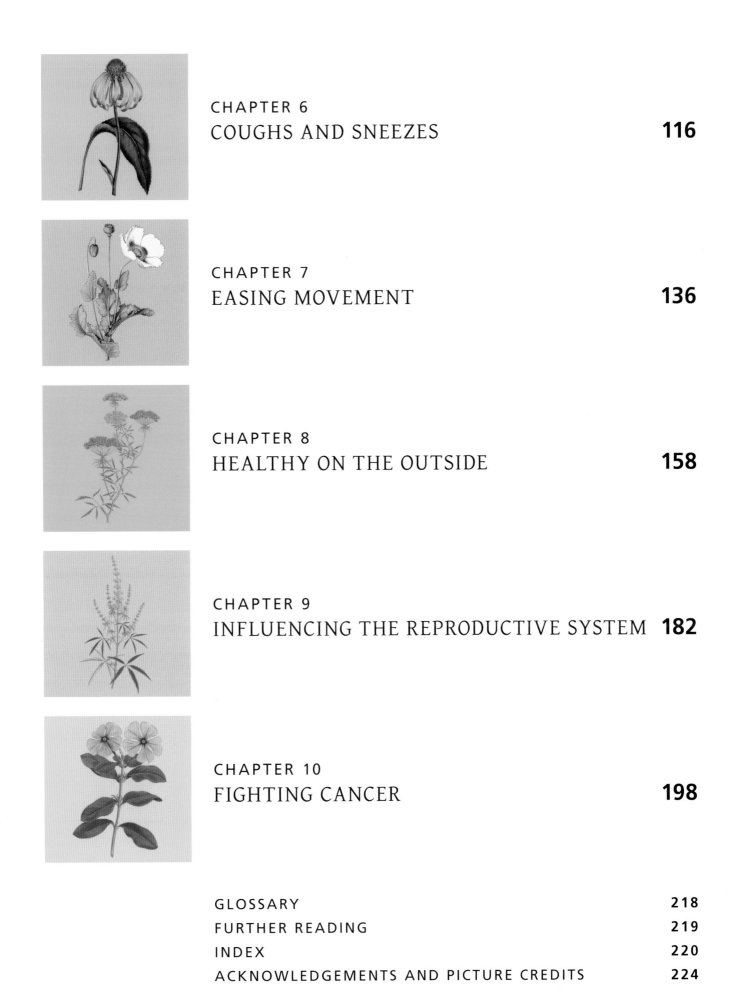

CHAPTER 6
COUGHS AND SNEEZES 116

CHAPTER 7
EASING MOVEMENT 136

CHAPTER 8
HEALTHY ON THE OUTSIDE 158

CHAPTER 9
INFLUENCING THE REPRODUCTIVE SYSTEM 182

CHAPTER 10
FIGHTING CANCER 198

GLOSSARY 218
FURTHER READING 219
INDEX 220
ACKNOWLEDGEMENTS AND PICTURE CREDITS 224

Introduction

Of the approximately 383,000 known plants worldwide, at least seven per cent have been used as traditional remedies. Modern science has proven the usefulness of many of these plants. This book introduces those plants from which isolated constituents have been developed as important pharmaceutical drugs, and other plants that are now widely available as herbal or dietary supplements with increasing evidence to support their use.

SCOPE

This book uses text, photographs, illustrations and chemical structures to explore plants and their constituents that humans have found useful as medicines. Chapter 1 explains how our use of plants globally as medicines developed from ancient times into various traditional systems – often with complex interactions, due to the movement of plants and medicinal knowledge around the world through trade, migration and exploration. Most fungi and other types of organism are outside the scope of this book, the exception being a few that exert their effects through their close association with plants.

The study of plants using scientific methods has enabled the discovery of an increasing number of active compounds. Some of these have been developed into pharmaceutical drugs, many of which will be familiar to the reader. Compounds may be used in their natural form or with modifications, or are fully synthesized using the plant compound as inspiration. The modern use of the compound may support the traditional use, but other plant compounds have been discovered simply through fortuitous chance. Some plant families are particularly rich in known medicinal plants, so can be explored as another approach in the search for medicinally useful compounds. The book introduces a few such families, giving some examples of their medicinal – as well as edible and poisonous – members.

BELOW **Purple foxglove (*Digitalis purpurea*) is one of the many plants examined in this book that has been used traditionally as a medicine and now provides us with important pharmaceutical drugs.**

HOW THIS BOOK WORKS

Title
Refers to the illness or condition covered on the pages.

Chemical structures
The structure of a naturally occurring active compound is illustrated for the featured plant if a single compound is particularly important.

Images
Photographs and botanical drawings to illustrate the plant or particular parts of the plant.

Sick of moving

Many of us will have experienced motion sickness when travelling by car, plane or boat, or on a fairground ride. In all these cases our body is accelerating in unfamiliar ways, and the visual cues don't match up with the messages relating to balance. One effective treatment currently available is hyoscine (also known as scopolamine), an alkaloid found in some plants in the potato family.

MAGIC IN THE BELLY

PLANT:
Hyoscyamus niger L.
COMMON NAME(S):
black henbane
FAMILY:
potato (Solanaceae)
ACTIVE COMPOUND(S):
NATURALLY OCCURRING:
tropane alkaloids (hyoscine [scopolamine], hyoscyamine)

SEMI-SYNTHESIZED:
hyoscine hydrobromide
MEDICINAL USES:
MAIN: motion sickness, to dry up secretions
OTHER: topical pain relief, fever, respiratory illness, nervous disorders
PARTS USED:
seeds, leaves, whole plant

ABOVE Hyoscine (scopolamine) is a tropane alkaloid found in certain members of the potato family (Solanaceae).

BELOW Black henbane (*Hyoscyamus niger*) is a biennial herbaceous plant with sticky, toothed leaves and cream flowers that usually have purple to brown venation. The appearance of its flowers and foul smell resemble rotting flesh and attract flies, which are its main pollinators.

Black henbane is the most widely distributed species in the genus *Hyoscyamus*, growing in temperate Eurasia and northwest Africa, and also the most commonly used in traditional medicine. The oldest medical text to mention henbane, the Ebers Papyrus, an Egyptian manuscript dating to 1500 BC, recommended it for 'magic in the belly' and was probably referring to Egyptian henbane (*Hyoscyamus muticus*).

Henbanes produce tropane alkaloids that in humans have anticholinergic (antimuscarinic) effects on the CNS and peripheral nervous system, with the types and proportions of alkaloids in the plant determining which effects are more pronounced. In studies, black henbane was found to have more sedative effects than belladonna, also known as deadly nightshade (*Atropa bella-donna*; see page 180), which also contains tropane alkaloids, and so it became the focus of research. In 1880, the German pharmacist Albert Ladenburg (1842–1911) isolated a compound he called hyoscine, which differed from the previously isolated tropane alkaloid atropine only by the addition of a single oxygen atom. At around the same time, the closely related Japanese henbane (*Scopolia japonica*) and Russian henbane (*S. carniolica*) were also found to contain tropane alkaloids, one of which was named scopolamine. In the early 1890s, hyoscine and scopolamine were discovered to be the same compound.

SUPRESSING SIGNALS

Hyoscine reduces the mismatch in visual and balance signals received by the brain – the cause of motion sickness – by suppressing nerve signals from the middle ear that are associated with balance. In addition, it facilitates the habituation processes by which the body becomes used to the movement. Hyoscine is taken (as a hydrobromide derivative) either in tablet form, or is applied as a transdermal patch, which exploits its ready absorption through the skin. Historic reference to this quality is seen in the inclusion of henbane and deadly nightshade in the ointments that witches are said to have applied to their skin and that led them to experience the sensation of flying.

There are a number of other medical applications for hyoscine based on its anticholinergic action of drying up secretions, including as a premedication for inhalant halothane anaesthetics. In palliative care it is given to reduce excessive respiratory secretions, and its spasmolytic action can relieve bowel colic and the pain associated with it (see also page 90).

LEFT Egyptian henbane (*Hyoscyamus muticus*) is native to a region extending from northeast Africa east to India. It has a higher tropane alkaloid content than black henbane (*H. niger*) and has been collected from the wild as a commercial source of these compounds.

Devil's breath

In addition to sedation, hyoscine can induce a lack of will and amnesia. From at least medieval times, this effect was exploited during childbirth, giving rise to the term 'twilight sleep'. For a brief period in the early twentieth century, the practice was popular in the United States. The administration of injections of hyoscine and morphine enabled women to wake up after the birth with no memory of the pain of labour. These women also became more suggestible, leading to the use of hyoscine as a 'truth drug'. In larger doses, hyoscine can be lethal, a fact that has been exploited by poisoners throughout history. Hawley Harvey Crippen (1862–1910), known as Dr Crippen, was one of the most notorious of these, using the drug to poison his wife.

ABOVE Hyoscine from angel's trumpets (*Brugmansia* spp.) is known as devil's breath in South America, where it is used by robbers to render their victims unconscious – many thousands of such incidents are reported each year in Colombia alone.

Fact file
Lists details of the featured plant(s), including: the scientific name and the author, with commonly used scientific names that are no longer current following in parentheses along with the designation 'syn.' for synonym; the plant family; the most widely used common names; the type(s) of medicinally active compounds that are 'naturally occurring' in the plant, with the most important compounds in parentheses; if relevant, the medicinal compounds that are 'semi-synthesized' using a naturally occurring compound as the starting material, or those that are 'synthesized from natural lead', having been inspired by a natural compound; medicinal uses divided into the main use, which is the subject of the spread, and the principal other uses; and lastly, the parts of the plant that are used medicinally.

Boxes
Used to look at the traditional uses of a plant or an item of more general interest.

ARRANGEMENT

After an introductory chapter, the other chapters of the book are organized by body system, with the first two pages of each chapter outlining some illnesses or conditions that can occur. Important plants from around the world that are used to treat these conditions or ease their symptoms are also introduced before being featured on subsequent pages. Many of these plants have a long history of use, and we tell their varied and intriguing stories. The active compounds and resulting pharmaceutical drugs are identified (where known or relevant), and their mechanisms of action on the human body and evidence for efficacy (effectiveness) are described. The latest research has been used throughout and is presented in a readily accessible way.

Coloured feature pages, interspersed throughout the book, introduce the floras and traditional systems of medicine from different geographic regions of the world. As with the species highlighted throughout the rest of the book, they demonstrate how vitally important plants are globally for our health and well-being.

Please note that this book is not a medical manual and is not intended as a guide to self-diagnosis and self-treatment.

INTRODUCTION

CHAPTER 1

THE BOTANICAL MEDICINE CHEST

The dependence of early humans on plants to provide medicines gave rise to a number of traditional systems of medicine. Scientific investigation of plants used traditionally has, in a number of cases, provided insights into the active compounds involved, their roles within the plant and their mechanisms of action on the human body. Some of these compounds have been developed into pharmaceutical drugs, and the modern classification of plants has further aided drug discovery.

St John's wort (*Hypericum perforatum*)

Medicinal beginnings

Until recently, humans have largely relied on plants to provide us with food, as well as materials for clothing, building and, of course, medicines. Which plants were useful as medicines was probably determined by trial and error. That long heritage has resulted in at least 28,000 plants being used in traditional medicine around the world and an estimated 80 per cent of the world's population still largely relying on plants for their healthcare needs.

ARCHAEOLOGICAL ANSWERS

It seems likely that as humans evolved we experimented with plants to treat our wounds, infections and diseases, just as we learnt which were nutritious and which were poisonous. Archaeology is providing some evidence of early use of medicinal plants. At a site in Israel through which several waves of early hominins (the immediate ancestors of humans and closely related species) are thought to have dispersed out of Africa, plant remains have been found, among other evidence of settlement. Dating to 780,000 years ago, during the Early to Middle Pleistocene, identified plants include some that are not edible and may have been used as medicines or fish poisons.

In a rock shelter in South Africa, bedding used by humans approximately 77,000 years ago has been preserved. It consists mainly of sedges and rushes, but there are also leaves from the Cape laurel (*Cryptocarya woodii*), a tree in the laurel family (Lauraceae). The choice of these particular leaves was probably no coincidence, as they contain volatile compounds that are effective against mosquitoes and other insects.

There is also evidence from the tooth enamel of our close relatives the Neanderthals that in Spain approximately 49,000 years ago they, too, used medicinal plants. These included poplars (*Populus* spp.) in the willow family (Salicaceae), which contain similar pain-relieving salicylates to willows (*Salix* spp.; see page 142).

WRITTEN WORKS

The traditional systems of medicine practised in China (see page 62) and India (see page 44) date back several thousand years, and some records from those times have survived. Civilizations that grew up around

LEFT **The castor oil plant (*Ricinus communis*) is native to the Horn of Africa but is widely naturalized in hot climates. It is cultivated as an ornamental, as well as commercially for its oil, which has both medicinal and industrial uses.**

3000–2000 BCE in Mesopotamia, an area now largely contained within Iraq and Syria, have left behind writing on clay tablets that includes descriptions of herbal medicines. Opium poppy (*Papaver somniferum*; see page 140) in the poppy family (Papaveraceae), common liquorice (*Glycyrrhiza glabra*; see page 88) in the legume family (Fabaceae) and willow are among the species that can be identified.

The Ancient Egyptians also kept medical records, on papyri. The best known of these is the Ebers Papyrus (*c.* 1552–1534 BCE), named after the German Egyptologist Georg Ebers (1837–1898), who bought the scroll in Luxor (Thebes). Hieratic text on the document lists 800 medicinal recipes, including 700 that contain medicinal plants. Although many remain unidentified, plants mentioned include liquorice; senna (*Senna alexandrina*; see page 92), another member of the legume family; and sea onion or sea squill (*Drimia maritima*; see page 28) in the asparagus family (Asparagaceae). Opium poppy was recommended for crying children, 'magic in the belly' was treated with Egyptian henbane (*Hyoscyamus muticus*; see page 86) in the potato family (Solanaceae), and the laxative and purgative properties of oil from the seeds of the castor oil plant (*Ricinus communis*) in the spurge family (Euphorbiaceae) are also described. Castor oil seeds found in Egyptian tombs dating to *c.* 4000 BCE provide evidence of the earlier use of this plant. Interestingly, papyrus itself is made from a plant – the pith of the aquatic papyrus sedge (*Cyperus papyrus*) in the sedge family (Cyperaceae).

ABOVE The bulbs of sea onion or sea squill (*Drimia maritima*) are the medicinal part of the plant. They grow near the surface of the soil and contain compounds that affect the heart. Sea onion produces tall spikes of white flowers after the leaves have died back.

Animal behaviour

Animals other than humans – including chimpanzees (*Pan troglodytes*) – have been observed eating plants not generally considered to be food, and it is thought that in these cases they are self-medicating. Plants eaten include some that are effective against intestinal parasites due to their high tannin content, and studies have shown that chimpanzees that eat these species do suffer from fewer parasites. It is highly likely that the use of medicinal plants has an evolutionary advantage, and scientists speculate that selection of medicinal plants was within the capabilities of the last common ancestor between chimpanzees and humans, which lived 6 million to 7 million years ago.

Traditional systems

Knowledge of plants was already well developed by the time the oldest surviving medical records were being written, and a rich exchange of people, ideas and material resulted in some exotic plants being incorporated into local traditions. Varying systems for classifying and treating diseases still exist today, and the best known are introduced here. The rich diversity of plants and medicines of nine geographic regions are discussed separately throughout the book.

ABOVE Ayurvedic medicine is a traditional system from India that commonly uses up to 300 plant species in its remedies.

RIGHT The ruins of the Asclepeion of Kos, in the Dodecanese, Greece. This temple was dedicated to Asclepius, the Greek god of medicine.

(*bhutas*), which are related to the five senses: earth (*prithvi*)/smell, water (*jala*)/taste, fire (*tejac*)/vision, air (*vaju*)/touch and space or the ether (*akasa*)/hearing. These five elements relate to the human body as three basic principles or humours, the *tridosha* (singular *dosha*), namely *vata*, *pitta* and *kapha*. Illness arises due to imbalance between the elements. Several thousand plants in India have medicinal applications, with 250–300 commonly used in Ayurvedic medicine, where they may be combined with minerals.

AYURVEDA

The traditional system of medicine practised most widely in India (see page 44) is known as Ayurveda, from *ayur*, meaning 'life', and *veda*, meaning 'knowledge'. Its roots are found in the Veda – a sacred Hindu text – and particularly the Atharva Veda (*c.* 1200 BCE), which provides a comprehensive description of medicine at the time. From around 600 BCE, Ayurvedic medicine was developed and a number of medical texts were written. The system is a combination of art, science and philosophy of life, and covers healthy living, personal and social hygiene, and the prevention and cure of diseases.

The living (including humans) and non-living environment is considered to consist of five elements

EUROPEAN MEDICINE

Indian and Egyptian beliefs and medicine influenced the work of ancient Greek physicians such as Hippocrates of Kos (*c.* 460–*c.* 370 BCE) and Galen of Pergamon (*c.* 129–200 CE). Four humours linked to vital fluids of the body were believed to correspond to the four elements that made up the world: blood/earth, phlegm/wind, black bile/fire and yellow bile/water. In addition to affecting health, they were also thought to have an impact on temperament: sanguine, phlegmatic, melancholic and choleric. Balance was restored by blood-letting, purging and herbs. The Greek physician and botanist Pedanius Dioscorides (*c.* 40–90 CE), who like Galen worked for the Roman army, wrote a comprehensive herbal that included more than 600 plants. His work, together with those of later Arab scholars (see page 84), influenced European herbal medicine for centuries (see page 190).

Doctrine of signatures

Many early systems of medicine around the world considered that the characteristics of a plant, such as its appearance, provided clues to its medicinal uses. The Chinese, for example, used the human-shaped roots of ginseng (*Panax ginseng*; see page 64) as a general tonic. In Europe, this belief can be seen in the works of Galen, Dioscorides and others, and was later revived by the Swiss alchemist Paracelsus (1493–1541). It was popularized subsequently by Jakob Böhme (1575–1624), a shoemaker in a small town in Germany who studied the works of many authors, including Paracelsus. Böhme's book *Signatura Rerum: The signature of all things* (1621) described a vision in which he had seen the relationship between God and humans. In addition to writing that God marked things with a sign or signature of their purpose, he noted that the sign might be seen in the environment in which the organism grew.

The popularity of the doctrine continued into at least the eighteenth century, before it was relegated to folklore.

ABOVE **Nutmeg (*Myristica fragrans*)** seed kernels. The appearance of nutmeg resembles the structure of the brain, hence its traditional use for conditions affecting that organ, including headaches and psychological disorders.

ABOVE **One of a series of woodcuts of illustrious physicians and legendary founders of Chinese medicine from an edition of *Ben Cao Meng Quan* (*Introduction to the Pharmacopoeia*), dating from the reign of the Ming dynasty Wanli Emperor (1573–1620).**

TRADITIONAL CHINESE MEDICINE

Inscriptions carved on animal bones dating to 5,000 years ago are the earliest preserved evidence of medicine in China. The first extant Chinese pharmacopoeia, written around *c.* 25–200 CE but incorporating a much earlier work, is the *Shen Nong Ben Cao Jing* (*The Drug Treatise of the Divine Countryman*, or *The Divine Farmer's Materia Medica Classic*), which includes 365 mainly plant-based drugs. It is part of a Taoist tradition and was used in combination with meditation, a special diet, exercise and other practices in order to achieve a long life. Emotional, spiritual and physical health are seen as necessary to well-being, and the practice of traditional Chinese medicine seeks to maintain balance and harmony in all these areas.

Several concepts underpin traditional Chinese medicine, including *qi* (pronounced 'chi'), which is the universal life force; the complementary opposites of *yin* and *yang*; and the five elements, *wu xing*, which correspond to the five vital organs in the human body, namely fire/heart, wood/liver, earth/spleen, metal/lungs and water/kidneys. Disease is seen as arising from a weakness in one or more of these organs, and can be due to external forces – the 'six excesses' – or the internal effects of the 'seven emotions'. Prescriptions personalized to individual patients usually contain five to 15 ingredients (mostly herbal). Traditionally, these are taken as decoctions, but manufactured single- or multi-ingredient products are commonly used today (see also page 62).

JAPANESE KAMPO

Traditional medicine in Japan (see page 62) is known as Kampo (or Kanpo), and although it has its roots in traditional Chinese medicine there are important differences. Kampo consists of *syou* (or *sho*), a diagnostic system in which each *syou* is a combination

ABOVE **Preparing traditional Japanese Kampo medicine prescriptions involves measuring out the herbs from drawers into individual trays (seen on the workbench at centre).**

of symptoms; and *houzai* (or *hozai*), the herbal medicines that are specific to the *syou*. Unlike traditional Chinese medicine, in which a practitioner will make up an individual prescription from the raw herbal drugs, the Kampo system mainly relies on a limited number of multi-herb formulae in specific doses. These are usually prescribed as ready-made manufactured preparations.

Plants continue to provide medicines around the world and many of the pharmaceutical drugs we rely on, and that are described in this book, were inspired by traditional use. Although the link between psychological and physical health – so often a part of traditional systems of medicine – is now receiving increasing acceptance, herbal preparations are far removed from the single-compound drugs usually preferred by conventional medicine (also known as Western, allopathic and mainstream medicine). In contrast, traditional medicine continues to use whole plants or plant parts, or even combinations of multiple herbs, albeit often as extractions in the form of tinctures, powders or tablets. Preparations will therefore include many hundreds or thousands of compounds that can have complementary (synergistic), opposite or additive effects.

The importance of names

Plant names can hinder the use of traditional systems of medicine, particularly in countries where they are not native. In traditional Chinese medicine, a single Pinyin name is often applied to herbal drugs sourced from more than one plant. Although usually members of the same genus, they can be from different genera and have different chemical profiles. Traditionally in the herbal trade the Chinese Pinyin name *fang ji* has been used to refer to the roots of either fourstamen stephania (*Stephania tetrandra*) in the moonseed family (Menispermaceae) or Chinese birthwort (*Aristolochia fangchi*) in the birthwort family (Aristolochiaceae). The latter contains toxic aristolochic acids, compounds that are harmful to the kidneys (nephrotoxic), and when misused can cause kidney disease. Inadvertent use of Chinese birthwort rather than fourstamen stephania by a slimming clinic in Europe in the 1990s led to kidney failure in more than a hundred patients. This highlights the need for careful identification of herbal material and unambiguous labelling on products.

ABOVE RIGHT **Roots of Chinese birthwort or guang fang ji (*Aristolochia fangchi*) can cause kidney disease due to the presence of aristolochic acids.**

RIGHT **Slices of *fang ji* (*Stephania tetrandra*) root are used in traditional Chinese medicine.**

Plant laboratories

Unlike animals and people, plants are fixed to the ground, so cannot escape from danger by running away and are unable to move to find a mate to reproduce. Plants have other survival strategies resulting from the selective effects of such environmental challenges. Strategies include the development of enzyme systems that enabled them to produce a diverse array of extremely complex chemicals with many different functions.

ABOVE **Passion flower or maypop (*Passiflora incarnata*) is a climbing plant with attractive flowers from central and eastern United States. The pulp of the ripe fruit is edible and was used by Native Americans for at least 3,000 years before the arrival of Europeans.**

MAKING A RAINBOW

Certain types of plant compounds give flowers, fruit and, sometimes, leaves a rainbow of different colours. These plant pigments do not just give us beautiful gardens – they have other functions, vital to the survival of plants. They give many flowers a palette of colours, which attract pollinators such as bees. This allows pollen, containing genetic material, to be transferred between flowers, aiding pollination. The bright pigments in fruit help to attract birds or mammals, which eat the fruit and so disperse the seeds. Both strategies aid the survival of plant species.

Plant chemicals, also referred to as compounds, that give yellow, orange and red colours include carotenoids, such as *beta*-carotene in carrots (*Daucus carota* subsp. *sativus*) from the carrot family (Apiaceae), and have been of interest for our health as part of our diet (see page 216). Some flavones and their derivatives may also be yellow; they occur in various medicinal plants, including St John's wort (*Hypericum perforatum*) in the St John's wort family (Hypericaceae; see page 60) and passion flower (*Passiflora incarnata*) in the passionfruit family (Passifloraceae). Both have been used traditionally for certain nervous conditions.

Anthocyanins give blue, purple and red colours, and occur in plants such as the flowers of roselle (*Hibiscus sabdariffa*) in the mallow family (Malvaceae; see page 31), used traditionally for its reputed diuretic effects and for cold symptoms. Other plant pigments include anthraquinones and their derivatives. These are often orange or red compounds, and occur in some medicinal plants such as those with laxative properties, including the pods of senna, frangula bark from alder buckthorn (*Frangula alnus*) in the buckthorn family (Rhamnaceae), and rhubarb (*Rheum palmatum*) in the knotweed or dock family (Polygonaceae).

WEATHER WITHSTANDERS

When plants, and people, are exposed to ultraviolet light from the sun, free radicals (also called reactive oxygen species) can be generated that react with cell components such as DNA and proteins, causing damage. Antioxidants can help counteract this damage. Many plant compounds – especially flavonoids and anthocyanins – have antioxidant properties and help protect plants from the harmful effects of the sun. Plant antioxidants from our diet are also of interest in helping to protect our cells and maintain our health (see also page 216).

Some plants are able to live in dry climates because they produce compounds that aid the preservation of moisture in conditions of drought. For example, aloe vera (*Aloe vera*) in the asphodel family (Asphodelaceae; see page 164) contains aloe gel in its leaves, which is rich in polysaccharides that form a mucilage to retain water. Other plants have strategies that enable them to survive in cold climates, including accumulating certain sugars and proteins (such as dehydrins) that help them tolerate freezing conditions.

TOXINS AND DETERRENTS

A number of plants produce toxic compounds that act as warning chemicals and poisons, to protect them from danger and prevent them from being eaten by predators. Many of these toxic compounds are alkaloids – for example, those that target the nervous system of insects. The potent biological effects of many alkaloids, and other compounds, have provided leads to discover useful pharmaceuticals, including those for heart disorders (see Chapter 2), nerve conditions (see Chapter 3) and cancer (see Chapter 10).

RIGHT **Flowering plants of roselle (*Hibiscus sabdariffa*), a shrubby annual or perennial from tropical West Africa and east to Sudan. After flowering, the fleshy red sepals expand and are used to make a drink, known as *jus de bissap* in Senegal and by other names elsewhere.**

Harnessing the power of plants

Some plants produce chemicals that provide protection from diseases caused by fungi or bacteria. Those in the mint family (Lamiaceae; see page 32), for example, produce essential (volatile) oils containing fragrant compounds, including monoterpenes. Some of these chemicals have antimicrobial properties, such as thymol from thyme (*Thymus vulgaris*) oil, which helps protect the plant from disease. Humans have learnt to harness these antiseptic powers – indeed, thymol has been used in antiseptic gargles and inhalant remedies for colds. Thyme oil itself has been of interest for potential use against head lice and scabies (see page 177).

THE BOTANICAL MEDICINE CHEST

A new era of discovery

Plants have been used for millennia for their medicinal properties. Traditional preparations were complex mixtures of many different plant compounds, as whole plant parts were the main source of medicines up until the end of the eighteenth century. In the early nineteenth century, however, a new era in the discovery of medicines began, changing the landscape of drug development and laying the foundations for the familiar conventional pharmaceuticals of today.

THE MEDICINES REVOLUTION

The discovery of drugs, or biologically active constituents, from plants or other natural resources is known as pharmacognosy, a term derived from the Greek *pharmakon*, meaning 'drug', and *gignosko*, meaning 'to acquire knowledge of'. The practice of pharmacognosy in the nineteenth century revolutionized drug discovery. In 1804, the pain-relieving chemical morphine was first isolated from opium, the latex of the opium poppy (see page 140). This marked a new era of isolating single chemicals from plants for use as medicines, rather than using mixtures of compounds in the form of herbal medicines.

As knowledge of chemistry and pharmacology became more advanced, drug discoveries from plants escalated. Other important drugs isolated from plants later in the nineteenth century included codeine, an additional analgesic from opium; the anti-malarial quinine from the bark of cinchona trees (*Cinchona* spp.) in the coffee family (Rubiaceae; see page 134); atropine for eye conditions from belladonna (*Atropa bella-donna*;

ABOVE **Polarized light micrograph of quinine sulfate crystals. Quinine was isolated from the bark of cinchona trees, including** *Cinchona officinalis* **(right), and was found to be useful in preventing and treating malaria.**

see page 180) in the potato family; the anaesthetic cocaine from the coca plant (*Erythroxylum coca*; see page 76) in the coca family (Erythroxylaceae); and cardiac glycosides for heart conditions, from foxgloves (*Digitalis* spp.; see page 38) in the speedwell family (Plantaginaceae). The chemical structures of these and many other plant compounds were determined, and methods to test their biological activities were also developed. Today, these pharmaceuticals remain important in our armoury against a wide range of diseases, and we have plants and pioneering early scientific research to thank for this.

FOLKLORE TO PHARMACY

The importance of traditional medicines in providing many common drugs that are still in current use cannot be overestimated. But how do scientists choose which plants to study in order to find new medicines? One approach is to select plants used in a particular traditional medicine system, and then study them to determine whether there is any scientific basis for their reputed medicinal effects – this is known as ethnopharmacology.

While this research field became more popular from the 1960s, scientific evaluation of medicinal plants was being used back in the nineteenth century to isolate drugs from plants. For example, the bark of willow (*Salix alba*) in the willow family was used for centuries for its reputed antipyretic, analgesic and anti-inflammatory properties. In the nineteenth century, the salicylates were first isolated from the bark, and science uncovered these as the active constituents. This discovery led to the synthesis of aspirin at the end of that century – a key drug for

various conditions and one that is still used in modern medicine (see pages 49, 142, 171 and 173). This is an early example of how active constituents in plants can be used as template chemical structures to design new and improved medicines. Other examples include the design of the drug rivastigmine, used for dementia symptoms, which was based on the alkaloid physostigmine from the Calabar bean (*Physostigma venenosum*; see page 72) in the legume family. The anticancer drug topotecan was designed from the chemical structure of the alkaloid camptothecin, from the bark of the happy tree (*Camptotheca acuminata*) in the tupelo family (Nyssaceae), which is used in traditional Chinese medicine for cancer (see page 208).

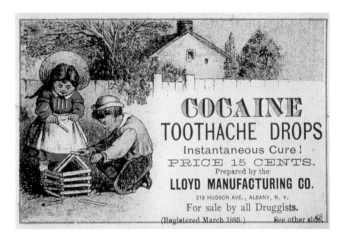

ABOVE Advertisement for cocaine toothache drops from 1885. Cocaine was isolated from the coca plant (*Erythroxylum coca*) in 1860 and became widely used as a local anaesthetic, but its use is now restricted.

A first in history

In 1864, the first edition of the British Pharmacopoeia was introduced, and although the monographs were dominated by crude herbal drugs, the tide was changing. This first edition described 'Morphiie Hydrochloras', referring to the 'hydrochlorate of an alkaloid prepared from opium', and 'three grains' of the latter were used to prepare 'Morphia Suppositories'. More than 150 years later, the British Pharmacopoeia still includes

monographs for morphine and its pharmaceutical preparations, along with many other plant-derived drugs.

LEFT Green glass bottle for 'LIQ: MORPH: HYD', or morphine hydrochloride, Europe. Morphine was isolated from the opium poppy (*Papaver somniferum*) in 1804 and is named for Morpheus, the Greek god of sleep.

Emerging technologies

In recent years, different approaches to finding new drugs have dominated medicines discovery. These include combinatorial chemistry, which involves synthesizing vast numbers of new chemical entities as potential drugs, and the emergence of biological agents such as antibodies, used for diseases such as cancer, multiple sclerosis and arthritis. Consequently, plants have been overshadowed as sources of pharmaceuticals, but their significance in past, present and future medicine cannot be forgotten.

ABOVE **In the 1970s, the East Asian Japanese plum yew (*Cephalotaxus harringtonia*) was found to contain a compound with anti-leukaemia activity, although it took 40 years for scientists to develop an approved drug.**

NEW DISCOVERIES

There are more than 383,000 known plant species, and much of this biological diversity remains unexplored in the search for new drugs. Recent drugs developed from plant sources include ingenol mebutate from petty spurge (*Euphorbia peplus*; see page 214) in the spurge family, and omacetaxine from the Japanese plum yew (*Cephalotaxus harringtonia*; see page 207) in the yew family (Taxaceae), which are both used in cancer therapeutics. Maytansine – originally sourced from certain plants in the spindle family (Celastraceae; see page 212) – may be produced by a bacterial community in the roots of these plants, and has joined technological advances in cancer therapeutics. Its analogue, DM1, has been combined with the antibody trastuzumab to give another new anticancer therapeutic agent, trastuzumab emtansine.

Many other plant compounds are under evaluation for their anticancer properties. They include betulinic acid, a triterpene from the bark of birch trees (*Betula* spp.) in the birch family (Betulaceae), which is being studied for its potential usefulness against conditions that may lead to skin cancer. Other plant-derived compounds are being assessed for their effectiveness against various diseases. For example, *scyllo*-cyclohexanehexol, which occurs in dogwoods (*Cornus* spp.) in the dogwood family (Cornaceae), and smilagenin from *Agave* spp. and *Yucca* spp., both in the asparagus family (see pages 28 and 162), have recently been tested for any potential they may have in therapeutics for Alzheimer's or Parkinson's disease.

BREWING UP PLANT DRUGS

Many plants remain vital as sources of pharmaceutical drugs because the complexity of the chemicals they produce is such that they cannot be synthesized in laboratory conditions. Indeed, we still rely on plants such as foxgloves for the heart drug digoxin, the opium poppy for the analgesic morphine and related alkaloids, and the Madagascar periwinkle (*Catharanthus roseus*; see page 210) in the dogbane family (Apocynaceae) for the anticancer drugs vincristine and vinblastine. However, our reliance on plant sources for these and other important drugs can put pressure on their survival.

Modern science has explored new ways to make these chemicals in order to meet the demands of the pharmaceutical industry, while also contributing to conservation strategies to preserve the biodiversity of our planet. In recent years, scientific research has provided a new understanding of the genes and pathways plants use to synthesize certain chemicals. For example, enzymes from plants and other organisms have been used to engineer new strains of yeast that can be 'brewed' to convert sugar into drugs that were originally discovered in plants. Drugs produced using this technique include opioid analgesics, cannabinoids (which are of interest for some nerve-related conditions) and the anti-malarial artemisinin. Although yields of these drugs are currently low using this method, the technology has great potential and avoids the need to harvest plant material to obtain them. The pathway for the synthesis of vinblastine in plants has also been uncovered, providing new insights into how this and other important anticancer drugs might be produced in other organisms. These strategies provide hope for increasing yields of these valuable medicines in the future.

New starring roles

The future for drug discovery from plants may also be in drug repurposing, where drugs used for one condition are tested for their potential efficacy against a completely different disease. This process enables the acceleration of drug approval by medicines regulators, as they have already been scrutinized for their effects in humans. For example, the anti-malarial drug artemisinin from sweet wormwood (*Artemisia annua*; see page 134) in the daisy family (Asteraceae) and aspirin, derived from chemicals originally discovered in willow bark, are being explored for their potential usefulness in cancer therapeutics.

LEFT Flowering dogwood (*Cornus florida*), whose native range is from southeast Canada to Mexico, contains a chemical under investigation for use in patients with Alzheimer's or Parkinson's disease.

BELOW A chemical constituent from silver birch (*Betula pendula*) has potential applications for skin cancer.

Medicines regulation

Legislation or regulation of medicinal plants varies between countries and depending on the form of the preparation. Pharmaceutical drugs are generally single compounds, while plant parts and extracts may be herbal remedies, botanicals, or dietary or food supplements. Plants with claimed health benefits can even be regulated as novel foods. Regulations in Europe and the United States are provided here as examples.

PHARMACEUTICAL DRUGS

The pages of this book chart the stories of many pharmaceutical drugs that have been isolated from plants or derived from plant compounds, either directly or by inspiration. Numerous plant compounds have been investigated as potential pharmaceutical drugs over the last century, but the vast majority fail at some stage in the long process. This process involves many stages, including chemical identification, mechanism of action and toxicity studies, and clinical trials. The few compounds that pass all these stages will receive a market authorization or product licence from a medicines regulator, such as the European Medicines Agency (EMA) or the United States Food and Drug Administration (FDA; this also regulates foods). The regulator will allow the supply of the compound as a pharmaceutical drug for specific purposes (indications), and all aspects of the product are tightly controlled to ensure it meets standards for quality, safety and efficacy (effectiveness). In all these respects, there is no difference between a pharmaceutical drug from, or inspired by, a plant, and one that resulted from chemical synthesis – such as, respectively, aspirin (see page 142) and paracetamol (acetaminophen). For reasons of historical use, some European countries (notably Germany) have also granted product licences to a number of plant products that might otherwise be considered herbal remedies.

HERBAL REMEDIES OR BOTANICALS

In Europe, a different set of regulations, introduced in 2004 and also administered by the EMA, enable the registration of products containing parts or extracts of plants. These can be sold over the counter as herbal remedies for minor self-limiting conditions, based on evidence for their traditional use. This evidence should relate to at least 30 years of use, including 15 years in Europe, and is in place of scientific evidence of efficacy from laboratory or clinical studies. Examples of some plant species that are included in such traditional herbal medicinal products can also be found in this book. They contain numerous and diverse chemical constituents, and while there may be some knowledge of which of these are medicinally 'active', there is no requirement to identify them or their mechanisms of action. Natural variation in the chemical constituents of individual plants can be addressed by ensuring a particular concentration of one of more chemicals in the final product. There are also requirements concerning the quality and safety of these products.

The United States allows only pharmaceutical drugs to make claims to treat, mitigate or prevent a disease. However, in 2004 it issued guidance for a 'botanicals' pipeline for drugs; by taking into account the history of traditional use of the product as a medicine, the clinical trials process can be shortened. These natural products can also be mixtures rather than single compounds. Although this process has been started for around 600 mixtures, only 2 per cent have made it to phase III clinical trials, and only two botanical drugs have been approved: crofelemer (see page 94) and sinecatechins (see page 172).

Supplementing your diet

Many countries allow products containing plant parts or extracts to be sold as foods without the need for authorization or registration. In the United States, these products – termed dietary supplements or nutraceuticals – are required to be 'generally recognized as safe', or having GRAS status, and should not make medicinal claims. Nutraceuticals are concentrated forms of foods, whereas dietary supplements may contain plant or other substances. In Europe, the European Food Safety Authority oversees the regulation of these products as food supplements intended to augment rather than replace a balanced diet. Health claims for food supplements require a certain level of scientific evidence, although many plant-based food supplements display health claims despite lacking authorization.

LEFT **The leaves of ginkgo (*Ginkgo biloba*; see page 74) are available in Europe both as traditional herbal remedies and food supplements.**

BELOW **Sinecatechins are a mixture of catechins from the green leaves of tea (*Camellia sinensis*) and are one of only two drugs that have so far been approved by the United States Food and Drug Administration via its 'botanicals' route.**

Botanical collections

Plants are a fundamental part of life on Earth, providing food and medicines, raw materials, and 'ecosystem services' such as stabilizing soil, cleaning air, and cooling the planet by trapping carbon dioxide and absorbing heat from the sun. A greater understanding of plants and their relatedness will enable us to protect them better while also revealing potential new sources of useful compounds or materials. Botanical collections are at the heart of this work.

ABOVE **The Herbarium at the Royal Botanic Gardens, Kew, in the United Kingdom, dates to 1853. Millions of specimens are now preserved in its cupboards (above) and are used by scientists (below) to undertake a range of research activities.**

NOT JUST GARDENS

It is likely that humans have always taken useful plants with them when they travel, and brought home plants when they explore new areas. Since the eighteenth century, expeditions have often been accompanied by botanists, who recorded the plants that were found and collected both living and dried specimens. These specimens made their way back to Europe and became the kernels of both private and public botanical collections. Since then, these original collections have grown and numerous others have arisen around the world, preserving examples of both national and exotic plants. Some of the earliest collections, such as those curated at the Royal Botanic Gardens, Kew, now contain many millions of dried plant specimens (herbarium specimens) and grow many thousands of living plants. In addition, there are collections of preserved seeds (in seed banks), published information, illustrations, and specimens relating to the use of plants (economic botany), their anatomical features (microscopy slides), their chemistry and their DNA.

CLASSIFICATION AND AUTHENTICATION

Scientists – particularly taxonomists and systematists – use botanical collections to undertake a variety of activities that benefit society and preserve the biodiversity of our planet. Underpinning all other activities is the ongoing discovery and identification of new plant species and the classification of plant groups. International collaborations enable specialists in particular plant groups to access collections held around the world. The use of specimens and an array of techniques – including anatomical, chemical and, particularly, DNA methods – have increasingly enabled plant classifications to reflect evolution. These classifications can indicate close relatives of medicinal plants that may be sources of the same or similar active compounds. This is useful if the original plant is rare or under threat, such as through overcollection for the medicinal trade or loss of the habitat where it grows.

Slightly different chemical structures are of interest as they might have better medicinal properties.

One important way in which botanical collections are being used is to enable the identity of medicinal plant material in the trade to be verified. This is known as authentication. Trade material is compared to specimens that have been expertly identified, including close relatives or likely substitutes. A number of botanic institutes also now have collections of verified 'voucher' herbarium specimens, which are linked to

plant material that has been processed according to traditional or regulated methods, including Chinese materia medica. Processing can affect the normal characteristics of a plant, so these collections can assist with authentication.

Conserving plants

As the human population expands, demands on the land and natural resources increase and are leading to a loss in biodiversity. This threatens not only the plants that become extinct, but the survival of all the life that depends on them. Botanical collections can help to quantify the loss and also protect this biodiversity. Collection labels on specimens include location details that can be used to map the distributions of plant species over time. Changes in distribution can be due to habitat loss, but they can also be caused by climate change, with species that are less able to adapt becoming scarcer than those that can. It is then possible to study the plants that are better able to survive to see which characteristics are key to this.

Collection details can also indicate areas of the world that should be prioritized for protection. They might be particularly species-rich, have high numbers of species that are threatened with extinction or include wild relatives of important crop plants. Botanical collections are used in the fight against illegal trade in plants, including plants collected from the wild for use in medicinal products. Shipments of plant material are checked when imported to see if they contain species that are listed by the Convention on International Trade in Species of Flora and Fauna, and botanical collections may be needed to assist with identifications. Lastly, seed banks are preserving the seeds of ever more species, and these can be germinated and grown on to provide plants, including for reintroduction to the wild and for research.

PLANT FAMILIES

Knowing the classification of a plant can help scientists predict which useful compounds it might produce. This is known as the chemosystematic or phylogenetic approach to drug discovery, as it uses knowledge of the relatedness of plants to look for species that might contain particular types of compounds. Some of the largest plant families that are important in traditional systems of medicine, and often also in conventional medicine, are introduced over the next few pages.

ASPARAGUS OR HYACINTH FAMILY

SCIENTIFIC NAME:
Asparagaceae

APPROXIMATE NUMBERS:
Species: 2,940 [1]

MEDICINAL PLANTS:
350 (12 per cent)

Numerous cardiac glycosides (a type of steroid) are widely found in plants and can be broadly divided into two types, the cardenolides and bufadienolides. The asparagus family is unusual in that it contains species that produce both sorts of these compounds. Cardenolides are the more common of the two and are found in species of lily-of-the-valley (*Convallaria* spp.; see page 39), while the less common bufadienolides occur in sea onion (see page 13) and related species. These plants and other members of the family – including butcher's broom (*Ruscus aculeatus*; see page 51), which contains steroidal saponins – have been used for heart or circulatory problems. Other plants such as sisal (*Agave sisalana*; see page 162) contain steroidal saponins that are used as corticosteroid and hormone precursors. Similar compounds are also found in species of asparagus (*Asparagus* spp.), including the one we eat as a vegetable – *Asparagus officinalis*. The specific epithet of this plant, *officinalis*, indicates its history of medicinal use; the rhizomes were used for kidney conditions. Other members of the genus are also used medicinally.

ABOVE **Young shoots or spears of asparagus (*Asparagus officinalis*) are eaten as a vegetable.**

LEFT **Leafy lily-of-the-valley (*Convallaria majalis*) shoots and heads of scented white flowers. This herbaceous perennial from Europe and the Caucasus has been used for heart conditions.**

[1] Bob Allkin et al. (2017). The average percentage of medicinal plants in a plant family is 8.3.

CARROT FAMILY

SCIENTIFIC NAME:
Apiaceae (syn. Umbelliferae)

APPROXIMATE NUMBERS:
Species: 4,080

MEDICINAL PLANTS:
590 (14.4 per cent)

ABOVE **Hemlock or poison hemlock (*Conium maculatum*), a biennial herb from Europe, North Africa and western Asia, was used for muscle spasms and the relief of pain.**

RIGHT **Parsley (*Petroselinum crispum*) is a Mediterranean herb with both culinary and medicinal uses.**

The 'seeds' (fruit) of many members of the carrot family produce essential oils that have reputed carminative properties and are traditional remedies for digestive complaints. They include dill (*Anethum graveolens*), caraway (*Carum carvi*), fennel (*Foeniculum vulgare*) and anise (*Pimpinella anisum*), and together with the fruit or leaves of other species such as coriander (*Coriandrum sativum*) and parsley (*Petroselinum crispum*) are valued as culinary herbs to flavour food – a way of eating your medicine (see also page 45).

The carrot family provides us with a number of vegetables, such as carrots, but also some of the most poisonous plants, including the hemlocks. In the 1860s, the piperidine alkaloid coniine, which is found in hemlock or poison hemlock (*Conium maculatum*), was the first plant alkaloid synthesized in a laboratory. Hemlock has been used medicinally as a spasmolytic and analgesic, but applications are now limited due to its toxicity. The poisonous water hemlock (*Cicuta virosa*), which contains the polyacetylene cicutoxin, had similar historical medicinal uses.

Furanocoumarins, such as the psoralens, are found in many members of the family, including giant hogweed (*Heracleum mantegazzianum*), notorious for causing severe skin reactions on contact, and also celery (*Apium graveolens*) and parsnip (*Pastinaca sativa*) plants. They are defence compounds, and are often produced in response to attack by fungal pathogens or at particular life stages, such as when seeds are setting. Their effects on skin have been harnessed therapeutically (see page 168). Furanochromones are structurally related to the furanocoumarins but are not photoactive (see page 124).

Another key medicinal plant in the carrot family is gotu kola (*Centella asiatica*; see page 166).

DOGBANE OR PERIWINKLE FAMILY

SCIENTIFIC NAME:

Apocynaceae

APPROXIMATE NUMBERS:

Species: 6,340

MEDICINAL PLANTS:

860 (13.5 per cent)

Cardiac glycosides of the cardenolide type are widely found in members of the dogbane family. Some have been used as ordeal or arrow poisons (see page 109), as well as medicinally (see page 39) for their effects on the heart. In southern Africa, the roots of uzara or milk bush (*Xysmalobium undulatum*) are used as a traditional remedy for diarrhoea. They contain uzarin-type cardiac glycosides that can have toxic effects.

The dogbane family is one of four in which indole alkaloids are mainly found. These compounds have a variety of medicinal applications, including for cardiac arrhythmias (see page 40), vascular dementia (see page 75) and the central nervous system (see page 68), and for cancer therapy (see page 209 and 210). Steroidal alkaloids are also occasionally found in the dogbane family. Conessi (*Holarrhena pubescens*, syn. *H. antidysenterica*) has been used traditionally for dysentery, as the name of one of its synonyms suggests. The active compound in the seeds and bark is the steroidal alkaloid conessine, which is active against a wide range of food-borne pathogens and also has anti-malarial properties.

LEFT **Conessi (*Holarrhena pubescens*) is a shrub or tree from parts of Asia and Africa. Both the seeds and bark are used medicinally.**

BELOW **Uzara or milk bush (*Xysmalobium undulatum* var. *undulatum*) is a herbaceous perennial with a natural range stretching from Ethiopia to South Africa.**

LEGUME, PEA OR BEAN FAMILY

SCIENTIFIC NAME:	MEDICINAL PLANTS:
Fabaceae (syn. Leguminosae)	2,330 (11.2 per cent)
APPROXIMATE NUMBERS:	
Species: 20,000–20,850	

Members of this large family include a number of medicinally important plants thanks to the array of active compounds they produce. The legume family is rich in alkaloids, including indole alkaloids (see pages 72 and 179), quinolizidine alkaloids (see pages 40 and 68), and guanidine derivatives in goat's rue (*Galega officinalis*; see pages 103 and 112). Coumarins are found in sweet clover (*Melilotus* spp.; see page 48) and furanocoumarins in fountain bush (*Cullen corylifolium*; see page 168). Dianthrone glycosides in senna (see page 92) produce strong laxatives, once the sugar (glycoside) part of the molecule has been removed by gut bacteria, and an anthrone derivative in araroba (*Vataireopsis araroba*; see page 169) was modified for use in psoriasis. Other medicinally useful compounds include triterpene saponins (see page 88), balsamic esters (see page 176) and isoflavonoids (see page 195). The amino acid levodopa (L-DOPA), found in some members of the family (see page 154), is being used to treat Parkinson's disease symptoms. Other legume species contain lectins, which are toxic plant proteins, and their effect of causing red blood cells to clump together has been used in blood typing.

MALLOW FAMILY

SCIENTIFIC NAME:	MEDICINAL PLANTS:
Malvaceae	620 (11.7 per cent)
APPROXIMATE NUMBERS:	
Species: 5,330	

The mallow family contains a number of plants that have medicinal applications but are probably better known for their other uses. Around 26 million tonnes of cotton are produced by cotton plants (*Gossypium* spp.; see page 188) each year, making this genus the most productive in the family. Delicious chocolate drinks, confectionary and desserts are made from the fermented beans (seeds) of cocoa (*Theobroma cacao*). They contain the stimulant alkaloids caffeine and theophylline; the latter is used medicinally around the world, including for respiratory conditions. Caffeine is also found in seeds of cola (*Cola nitida* and *C. acuminata*), used to flavour soft drinks, as well as in species from a number of other families (see page 66). Other members of the mallow family produce organic acids, such as the baobab tree (*Adansonia digitata*), whose fruit has become a popular health food; or polysaccharides with soothing properties, including mallow (*Malva sylvestris*), hibiscus (*Hibiscus* spp.; see also page 19) and the lime or linden tree (*Tilia cordata*).

ABOVE **Fountain bush (*Cullen corylifolium*) seeds contain defensive chemicals that are used medicinally. The seeds of many protein-rich legumes are eaten, but may require heating to destroy the toxins.**

ABOVE **Cocoa (*Theobroma cacao*) tree with ripening pods (fruit). In addition to caffeine and theophylline, the beans (seeds) contain the bitter-tasting alkaloid theobromine.**

MINT FAMILY

SCIENTIFIC NAME:

Lamiaceae (syn. Labiatae)

APPROXIMATE NUMBERS:

Species: 7,760

MEDICINAL PLANTS:

1,060 (13.7 per cent)

The mint family is valued for its essential (volatile) oils, which are used as food flavourings and perfumes. For the plant, however, the value of these compounds – which include monoterpenes – lies in their antimicrobial properties (see page 19), an action also harnessed by humans in medicines and household disinfectants. Many members of the family are discussed in this book for the roles their volatile oils and other constituents play in aiding sleep (see page 59), improving memory (see page 75), calming irritable bowel syndrome (see page 90), relieving pain (see page 145) and killing skin parasites (see page 177). Diterpenes contribute to the medicinal properties of some species, such as forskolin in the roots of forskohlii (*Coleus barbatus*, syn. *Plectranthus barbatus*, also incorrectly referred to as *Coleus forskohlii*), which is used in Ayurvedic medicine and has shown a plethora of biological activities. Other diterpenes have been found to modulate the response of cells to hormones (see also page 192).

LEFT AND RIGHT **Flowering plants of the larger variety of forskohlii or Indian coleus (*Coleus barbatus* var. *grandis*), a perennial herb whose tuberous roots are used in Ayurvedic medicine and whose leaves are used in Brazil for various conditions.**

POTATO OR NIGHTSHADE FAMILY

SCIENTIFIC NAME:
Solanaceae
APPROXIMATE NUMBERS:
Species: 2,600

MEDICINAL PLANTS:
350 (13.6 per cent)

Members of the potato family produce a range of alkaloids. Steroidal alkaloids and their glycosides are found in *Solanum* spp., a large genus that includes the staple root crop potato (*S. tuberosum*), and fruit crops such as tomato (*S. lycopersicum*; see page 216) and aubergine (*S. melongena*). These and other species in the genus have medicinal applications (for example, see page 215). Steroidal alkaloids are also found in a few other families, but the addictive nicotine, which is also an alkaloid, is less common and is found in the greatest amounts in the tobacco genus (*Nicotiana* spp.; see page 68).

The tropane alkaloids are largely restricted to the potato family, an exception being cocaine in the coca plant (*Erythroxylum coca*; see page 76) in the coca family. They are probably the most medicinally important compounds in the potato family and are found in a number of different genera. The chemical structures of some tropane alkaloids have similarities to the neurotransmitter acetylcholine, so they can bind to muscarinic receptors in the central and peripheral nervous systems, and disrupt the transmission of particular nerve signals – they are said to be anticholinergic or antimuscarinic. The effects of each tropane alkaloid on the body can differ, depending on its affinity for certain receptors in the nervous system (see pages 80, 86, 138, 156, 161 and 180).

Other key medicinal plants in the potato family include chilli pepper (*Capsicum annuum*; see page 144) and Indian ginseng (*Withania somnifera*; see page 65).

SPURGE FAMILY

SCIENTIFIC NAME:
Euphorbiaceae
APPROXIMATE NUMBERS:
Species: 6,050

MEDICINAL PLANTS:
860 (13.5 per cent)

Diterpene phorbol esters, widely found in the spurge family (and also in the mezereum family, Thymelaeaceae), are often highly irritant and have been used traditionally to make arrow and fish poisons. Phorbol esters of the tigliane type found in the seed oil of some species of the genus *Croton* spp. are highly purgative, causing drastic diarrhoea (see page 133), while a few species of the same genus contain proanthocyanidins that may have anti-diarrhoeal effects (see page 94). Ingenol mebutate, which is an ingenol-type phorbol ester in petty spurge (see page 214), is used externally for skin keratosis, but not all phorbol esters are irritant (see page 133). A dose of castor oil, from the seeds of castor oil plant, is purgative due to the presence of ricinoleic acid; it is now largely used externally as a carrier oil (see page 177), as well as for numerous industrial applications. The roots of cassava (*Manihot esculenta*) are a staple food for 700 million people, despite the presence of the cyanogenic glycoside linamarin – the roots must be processed before they are safer to eat.

ABOVE Cassava (*Manihot esculenta*) is a woody shrub from South America that produces large clusters of toxin-laced tuberous roots.

LEFT The fruits of the tomato (*Solanum lycopersicum*) are a dietary source of lycopene.

CHAPTER 2

STRONG
AT HEART

Our heart is vital for our survival, maintaining the circulation of blood around the body. Diseases of the heart and circulatory system can affect the heart's rhythm or function, or the blood vessels. This chapter describes plants that have been instrumental in providing key pharmaceuticals to manage heart and blood circulation disorders. Examples of herbal medicines that have been of interest because of their traditional uses for the heart and circulation are also described.

Purple foxglove (*Digitalis purpurea*)

NATURAL COMPOUNDS FOR THE CIRCULATORY SYSTEM

Heart disease kills almost 18 million people worldwide every year and accounts for nearly one-third of all deaths. Four out of five deaths caused by heart disease are due to heart attacks and strokes, which affect the blood supply to the heart and brain, respectively. However, heart disease may be managed by pharmaceuticals, including important drugs originally derived from plant sources.

ABOVE **The flowers of purple foxglove (*Digitalis purpurea*) are large enough to enable pollination by bumblebees. Flower development and the pattern of nectar rewards encourage outcrossing.**

A STABLE RHYTHM

Some useful drugs may never have been discovered without scientific research on the chemistry and medicinal properties of plants. This is because some plant chemicals are so complex that they cannot be made from scratch in the laboratory. Indeed, plants are such brilliant chemists that they even outperform humans in their ability to synthesize certain chemicals. For example, digoxin, from the woolly foxglove (*Digitalis lanata*; see page 38) in the speedwell family (Plantaginaceae), is used as a pharmaceutical for certain heart disorders, including dysfunction of the heart rhythm. It has such a complicated chemical structure that it still has to be sourced from the plant to meet the demands of the pharmaceutical industry.

Other plant chemicals that have been developed as medicines for an irregular heart rhythm include alkaloids from Indian snakeroot (*Rauvolfia serpentina*) in the dogbane family (Apocynaceae), and from species of *Cinchona* trees in the coffee family (Rubiaceae). Khella (*Visnaga daucoides*), a plant in the carrot family (Apiaceae), contains the chemical khellin, which was the basis for the discovery of the anti-arrhythmic drug amiodarone (see page 41).

PRESSURE CONTROL

High blood pressure can be managed by different strategies, including certain lifestyle changes and pharmaceutical drugs that work in different ways. Current drugs used to manage high blood pressure include verapamil, which was designed from a chemical in the opium poppy (*Papaver somniferum*; see pages 46 and 140), which is in the poppy family (Papaveraceae). Other plant chemicals that have sometimes been used for high blood pressure include reserpine from Indian snakeroot and protoveratrines A and B from white hellebore (*Veratrum album*) in the trillium family (Melanthiaceae) (see page 46).

IN YOUR BLOOD

In some forms of heart disease, clots may develop in the blood vessels and can impair blood circulation, including the blood flow to vital organs such as the heart, brain and lungs. To help reduce the risk of clots, pharmaceuticals have been developed to 'thin' the blood. The discoveries of key drugs used for this purpose have particularly interesting stories in the history of medicine. The toxic effects in cattle of mouldy hay made from some species of sweet clover (*Melilotus* spp.) including yellow sweet clover or melilot (*Melilotus officinalis*; see page 48) in the legume family (Fabaceae) led to the discovery of a rat poison, and eventually to the anticoagulant drug warfarin. Another blood 'thinner' is aspirin, derived from chemicals in the bark of willow (*Salix alba*) in the willow family (Salicaceae), whose story of discovery began during a riverside walk in Oxfordshire in the eighteenth century (see pages 49 and 142).

HERBAL HISTORY AND THE HEART

In traditional herbal medicine, many plants have been used for their reputed effects to manage heart disorders, and some of these have been the subject of modern scientific studies to explore whether there is any rational basis for their traditional uses. Plants used traditionally for heart and circulatory disorders include hawthorn (*Crataegus monogyna*; see page 42) in the rose family (Rosaceae), horse chestnut (*Aesculus* hippocastanum) in the soapberry family (Sapindaceae) and butcher's broom (*Ruscus aculeatus*) in the asparagus family (Asparagaceae) (see page 50). Plants used for both their medicinal and culinary uses have also been of interest for any benefits they may have in managing heart disease; examples include garlic (*Allium sativum*; see page 49) in the amaryllis family (Amaryllidaceae).

ABOVE **Somali arrow poison (*Acokanthera schimperi*; see page 39) is a shrub or small tree native to the Democratic Republic of Congo, central east Africa and Yemen.**

ABOVE **Freshly dug roots of the herbaceous perennial Chinese sage (*Salvia miltiorrhiza*; see page 43), which is native to central and eastern China and Japan.**

In a heartbeat

Our heart function can be affected by diseases that change the normal regular rhythm of our heartbeat (arrhythmias), or that cause an inefficiency in how the heart muscle contracts. Pharmaceutical interventions are used to manage these conditions and include drugs derived from certain species of foxglove (*Digitalis* spp.) and Indian snakeroot. Other plants such as hawthorn have been used as traditional herbal medicines for heart conditions, although the scientific evidence to support their use has not been adequately substantiated.

WOOLLY FOXGLOVES

PLANT:
Digitalis lanata Ehrh.
COMMON NAME(S):
woolly foxglove, Grecian foxglove
FAMILY:
speedwell or plantain (Plantaginaceae)
ACTIVE COMPOUND(S):
NATURALLY OCCURRING:
cardiac glycosides (lanatosides, digoxin)

MEDICINAL USES:
MAIN: cardiac arrhythmias, heart failure
OTHER: externally for skin conditions
PARTS USED:
leaves

BELOW The cardiac glycoside digoxin was originally sourced from woolly foxglove (*Digitalis lanata*) and is still used pharmaceutically for certain heart conditions.

Digoxin

BELOW Woolly foxglove (*Digitalis lanata*) flowering plants.

The woolly foxglove is a biennial or perennial plant, native to southeast Europe and northwest Turkey, although it is cultivated in other parts of the world. The leaves contain different cardiac glycoside constituents, with the 'cardiac' part of their name indicating their effects on the heart. These cardioactive constituents include lanatosides A, B and C. Lanatoside C has been used alone, or in combination with other lanatosides, as a medicine for particular heart conditions. It is also converted into another cardiac glycoside, digoxin, which is one of the principal pharmaceuticals used in modern medicine for heart failure and certain types of arrhythmias. Digoxin was first sourced from woolly foxglove in the 1930s and today it is still derived from this plant for pharmaceutical use. Cardiac glycosides such as digoxin and the lanatosides increase the force of heart contractions, and they reduce the conduction of electrical impulses in the heart. These actions can help to regulate the heart rhythm and reduce symptoms in heart failure, including oedema.

Cardiac glycosides are also found in other species of *Digitalis*, including the purple foxglove (*Digitalis purpurea*), which is native to parts of Europe, including Britain and Ireland. Despite its common and species names, the flowers may be purple, pink or white. Constituents include purpurea glycosides A and B in fresh leaves, but when these are dried, the compounds are converted by enzymes to give digitoxin and gitoxin. Digitoxin is a potent cardiac glycoside that has also been used medicinally for the management of the same heart conditions as digoxin.

ABOVE **Climbing oleander (*Strophanthus gratus*)** has a native range from Senegal east to the Democratic Republic of Congo. Nomadic people in the centre of this region have used the seeds as the source of an arrow poison.

CARDIAC CHEMICALS

Cardiac glycosides have been found in other plants across different families in the plant kingdom. For example, the seeds of climbing oleander (*Strophanthus gratus*) and wood from the Somali arrow poison tree (*Acokanthera schimperi*), both in the dogbane family and both native to parts of Africa, are sources of cardiac glycosides, including ouabain. This has similar properties to digoxin, and so has also been used in the management of heart failure. In the same family is the East African kombe (*Strophanthus kombe*), whose seeds were formerly prepared as a tincture and used for particular heart conditions as they contain cardiac glycosides based on the chemical strophanthidin.

Lily-of-the-valley (*Convallaria majalis*), in the asparagus family, is native to parts of Europe, including Britain, France and Germany. Its rhizomes and aerial parts are another source of strophanthidin-derived cardiac glycosides. This plant has been used traditionally in herbal medicine for its cardioactive properties, but it was also reputed to strengthen memory and, according to folklore, was once used as an ingredient of love potions.

Fairy flowers

According to legend, foxgloves have many associations with fairies, who are said to dance under their flowers – hence the plants have been known as fairies' gloves, hats and dresses. Foxgloves were also believed to cure any disease caused by fairies or their spells, and were used in Ireland as a charm against witchcraft.

The poisonous effects of foxgloves have long been recognized, explaining why the plants have also been known as dead man's bells. Yet their medicinal properties have been valued throughout history. Traditionally, foxglove leaves were used externally for wounds, leg ulcers and eczema. In 1785, the British physician William Withering (1741–1799) documented the use of foxgloves for oedema, at a time when it was not realized this could be the result of a heart condition. More than 200 years later, with advances in medicine and new understanding of diseases, foxglove cardiac glycosides are still widely prescribed for heart conditions.

THE RHYTHM OF THE ROOT

PLANT:

Rauvolfia serpentina (L.)
Benth. ex Kurz

COMMON NAME(S):

Indian snakeroot,
sarpagandha

FAMILY:

dogbane (Apocynaceae)

ACTIVE COMPOUND(S):

NATURALLY OCCURRING:

indole alkaloids (ajmaline)

MEDICINAL USES:

MAIN: cardiac arrhythmias

OTHER: nervous disorders

PARTS USED:

root

RIGHT **The alkaloid ajmaline from the Indian snakeroot (*Rauvolfia serpentina*) has been developed as a pharmaceutical for certain types of heart arrhythmias.**

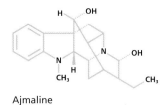

Ajmaline

ABOVE **Flowering branch of the Indian snakeroot (*Rauvolfia serpentina*) shrub.**

A high proportion of pharmaceutical drugs derived from plants are based on alkaloids, so it is not surprising that different alkaloids from Indian snakeroot have potent biological activities, revealed to be of medicinal importance. Discoveries about Indian snakeroot have not only played an important role in revolutionizing treatment strategies for depression (see page 61), but have also yielded drugs for both hypertension (see page 47) and cardiac arrhythmias.

Indian snakeroot has a native range extending from the Indian subcontinent through to south-central China. The root of the plant is known to contain at least 40 alkaloid constituents and one of these, ajmaline, has been developed as a pharmaceutical for certain types of cardiac arrhythmias and has been marketed in Japan for this purpose. The roots of the related African serpentwood or rauvolfia (*Rauvolfia vomitoria*), native to tropical West Africa, also contains alkaloids of medicinal interest, including reserpine, yohimbine (see page 196) and ajmaline. The chemical structure of ajmaline has been used to inspire the design of a new drug, lorajmine, also prescribed for the management of cardiac arrhythmias.

SYMPTOMS SWEPT AWAY

Broom (*Cytisus scoparius*), in the legume family, is a perennial shrub native to Europe. As indicated by the plant's common name, the stalk was indeed once used as a broom for sweeping and, according to legend, also by witches. Traditionally, broom had various medicinal uses. It was considered to be an antidote to poisons, and it was believed that sheep grazing on broom would be immune to snakebites. Broom was also reputed to have diuretic properties and to alleviate oedema, giving early clues that it might yield chemicals useful for heart conditions. Other applications included as a remedy for jaundice, a liver condition that may result in a yellowing of the skin. Broom was perhaps used in this way because its flowers are yellow and, according to the ancient doctrine of signatures (see page 15), plants that resemble parts of the body afflicted by a disease may be used as a remedy for that disease.

Broom was also once used traditionally to induce labour. This use can now be explained scientifically by the presence of the quinolizidine alkaloid sparteine in the leaves, which causes contraction of the uterus. Sparteine has a similar action on heart rhythm to that of quinidine (see box), and so has been used in some circumstances as a drug for particular types of cardiac arrhythmias.

ABOVE **Flowering broom (*Cytisus scoparius*) plant.**

At the heart of history

In 1630, the Countess of Chinchon, wife of a viceroy of Peru, was said to be the inspiration for the name of the trees in the genus *Cinchona*, whose bark gained a reputation at the time as a fever remedy. The bark was later discovered to contain the quinoline alkaloid quinine. Today, the bitter taste of quinine can be detected in tonic water, which is often combined with gin as a popular drink. Historically, quinine had a major impact as a treatment for malaria and it is still used in the fight against this parasite (see page 135). A related quinoline alkaloid in *Cinchona* bark is quinidine, which has antimalarial properties but can also change the conduction of electrical impulses involved in heart rhythm. For this reason, quinidine has been used therapeutically to manage certain types of arrythmias. It has sometimes also been used as an antimalarial drug.

A later discovery in the history of heart drugs occurred in the 1960s. Khellin is a furanochromone constituent derived from the fruit of khella and has been used for asthma (see page 124), but because it also dilates blood vessels, it has been used for angina. Khellin's chemical structure provided a template for the design and synthesis of new drugs intended to manage angina symptoms. One of these, amiodarone, was discovered to have anti-arrhythmic properties, so it was developed as a medicine for some forms of arrhythmia.

ABOVE **Tonic water fluorescing under ultraviolet radiation due to the presence of quinine.**

HEDGEROW HEALER

PLANT:

Crataegus monogyna Jacq.

COMMON NAME(S):

hawthorn, mayflower

FAMILY:

rose (Rosaceae)

ACTIVE COMPOUND(S):

NATURALLY OCCURRING:

flavones (including vitexin), flavonols (including

quercetin), procyanidins (including those composed of epicatechin and catechin units)

MEDICINAL USES:

MAIN: heart conditions

OTHER: nervous conditions

PARTS USED:

fruit, leaves, flowers

Hawthorn has a native range stretching from Europe through to the Caucasus and northern Africa. In Britain, it is common in hedgerows – so much so, in fact, that the herbalist Nicholas Culpeper (1616–1654) did not bother to describe it in his 1653 herbal, stating instead, 'It is not my intention to trouble you with a description of this tree, which is so well known that it needs none.' Culpeper did, however, record the traditional medicinal uses of hawthorn (see box).

In herbal medicine, different hawthorn species (especially *Crataegus monogyna* and *C. laevigata*) have long been used for heart conditions, and the berries (fruit), leaves and flowers have been studied to understand any scientific basis for such uses, although the actual species investigated have not always been defined. Hawthorn extracts prepared from either leaves, flowers or berries are considered to act by polyvalency; that is, different constituents – especially the flavonoids and procyanidins – work together to produce the overall biological effects.

In laboratory studies, hawthorn berry extracts have shown an ability to strengthen heart function and blood circulation, and to dilate blood vessels and lower blood pressure. Berry extracts have been tested in clinical trials in people with heart complaints such as heart failure, and although some studies suggest they may have some effect in these conditions, any significant benefits overall have not been convincing.

Hawthorn leaf and flower extracts have also been scientifically evaluated for any usefulness they may have in heart conditions. Whole extracts and purified extracts containing either concentrated flavonoid or procyanidin constituents have been shown to increase both the force of heart contraction and blood flow to the heart. Other laboratory studies reveal these different extracts may relax blood vessels and reduce blood pressure, and might have some ability to control arrhythmias. Leaf and flower extracts have been tested in humans with heart failure or other heart conditions in controlled clinical studies. While some studies suggested these extracts may improve symptoms, others showed no such benefits compared to a placebo. Therefore, there is insufficient evidence to date to support the use of hawthorn leaf, flower or berry extracts for heart conditions.

ABOVE **Fruiting hawthorn (*Crataegus monogyna*).**

100μm

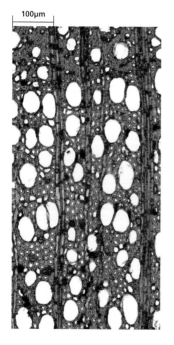

100μm

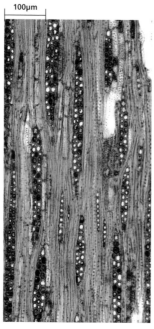

LEFT **Hawthorn (*Crataegus monogyna*) wood microscopy images stained red with safranin (transverse section on left and tangential section on right). Examining such images is part of the process of confirming that plant material in the medicinal trade is correctly identified.**

KEEPING IT IN THE FAMILY

Motherwort (*Leonurus cardiaca*), also known as lion's tail, is in the mint family (Lamiaceae) and is native to Europe and Iran. The aerial parts of the plant, including the flowers, have been used traditionally for their reputed ability to alleviate nervous disorders, menstrual problems and heart conditions. Laboratory studies suggest motherwort extracts do have cardioactive properties and may lower blood pressure, but more research is needed to investigate the biological effects of this plant and its active constituents.

Another plant used traditionally for heart conditions is Chinese sage (*Salvia miltiorrhiza*), also in the mint family. The roots are a traditional Chinese medicine considered to alleviate heart pain. Root extracts and their diterpenoid constituents have been tested in both laboratory studies and in humans for any usefulness they may have in heart conditions.

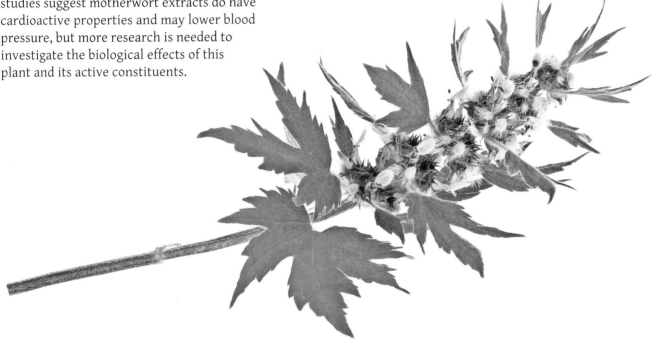

ABOVE **Motherwort (*Leonurus cardiaca*) is a herbaceous perennial that can reach 120 cm (47 in) tall. It bears dense whorls of white or pale pink flowers.**

ABOVE **Chinese sage (*Salvia miltiorrhiza*) flowers are purple-blue or white.**

Treating the torment

Across Europe and in other parts of the world where hawthorn grows wild, the berries, leaves and flowers of the plant have been used traditionally for various heart conditions. Indeed, the English herbalist Nicholas Culpeper claimed the powdered seeds drunk in wine were good 'for the dropsy', which is an old term for oedema and is now known to be associated with heart failure. Culpeper also used the seeds in wine 'for inward tormenting pains', suggesting another traditional use that may be related to the heart and perhaps referring to angina. In Ireland, hawthorn bark had a very different traditional use, as it was chewed as a remedy for toothache.

INDIA

India is a country of extremes. It is bordered in the north by the towering Himalayas, from where water drains south onto the Indo-Gangetic Plain. Arid and semi-arid conditions in the northwest of the country give way to increasing rainfall to the east and along the southwest coast. With such diversity, it is no surprise that India is home to at least 22,000 plant species, including 18,500 flowering plants, and that around 3,000 are used medicinally.

Ayurveda, Siddha and Unani

Ayurveda, the main system of medicine in India, is practised by at least 80 per cent of the population, which currently stands at around 1.37 billion. It is thought to be one of the earliest, if not the earliest, systems of medicine and has influenced others, including traditional European medicine (see page 14). Ayurveda uses medicinal plants and lifestyle guidance to promote health and long life, as well as to treat disease, but it isn't the only traditional system of medicine in India. Siddha medicine is said to have been described by the Hindu deity Lord Shiva. Its practice is largely restricted to the Tamil Nadu region of southeast India, as well as to Sri Lanka, and involves the use of herbs and minerals, along with yoga and fasting. The third main system of medicine, Unani (also known as Unani-Tibb), is a traditional Islamic practice that is followed today in central Asia (see page 84).

BELOW **Twelfth- to thirteenth-century copy of the *Sushruta-Samhita* (*A Treatise on Ayurvedic Medicine*), one of the core Ayurvedic texts, written in ancient Sanskrit on palm leaves.**

The Ministry of Ayurveda, Yoga and Naturopathy, Unani, Siddha and Homoeopathy regulates traditional medicine in India (homeopathy was introduced by the British during the British Raj, 1858–1947). Ayurvedic medicine is practised in 1,500 hospitals and there are more than 300,000 practitioners. A number of classical texts are the primary sources for the study of Ayurveda, together with commentaries and translations. In addition, there is an Ayurvedic Pharmacopoeia of India, the first volume of which was published in 1978. Similar combinations of classical texts and official pharmacopoeias exist for the Siddha and Unani systems of medicine.

Rasayanas

In Ayurvedic medicine, the *rasayanas* (literally rasa, meaning 'in the essence', and *ayana*, meaning 'what enters') are rejuvenating tonics that are believed to penetrate and revitalize the essence of a person's psycho-physiological being. Treatment with *rasayana* herbs is accompanied by mantras and meditation. Different *rasayanas* are prescribed for the three elements of the *tridosha* (see page 14). Ashwagandha or Indian ginseng (*Withania somnifera*; see page 65), in the potato family (Solanaceae), is prescribed for *vata*, and is the main male rejuvenator. For *pitta* you might receive the roots of shatavari or asparagus fern (*Asparagus racemosus*) in the asparagus family. Shatavari is also known as 'who

possesses a hundred husbands' and is the main female rejuvenator. Bibhitaki or beleric/belleric myrobalan (*Terminalia bellirica*), in the bushwillow family (Combretaceae), is one of the *kapha* tonics. It is strongly laxative and is used traditionally for cleansing the bowel, including to remove parasites.

ABOVE **The bibhitaki or beleric myrobalan tree (*Terminalia bellirica*) can reach 35 m (115 ft) in height. Its native range extends from India to western China and Malaysia. The dried fruit (pericarp) are used in all three systems of Indian medicine and traditional Chinese medicine.**

SPICE OF LIFE

Indian cooking uses a number of herbs and spices that not only provide aroma, taste and colour, but often also have medicinal properties. Many are known to aid digestion, such as the 'seeds' (fruit) of members of the carrot family, including cumin (*Cuminum cyminum*). The

ginger family (Zingiberaceae) also provides a few key flavours: the rhizomes of ginger (*Zingiber officinale*; see page 82) and turmeric (*Curcuma longa*; see pages 107 and 216), and also the 'pods' (fruit) or seeds of cardamom (*Elettaria cardamomum*). Cardamom pods are known as the 'queen of spices', and their use dates to the fourth century BCE. The 'king of spices' is black pepper (*Piper nigrum*) in the pepper family (Piperaceae). Piperine from black pepper has been studied for its potential for use in the skin condition vitiligo. The bark of true or Ceylon cinnamon (*Cinnamomum verum*) in the laurel family (Lauraceae) is used to ease some symptoms of irritable bowel syndrome and has also been studied for use in diabetes. Although native to Sri Lanka, it has been introduced to India.

LEFT **Cardamom (*Elettaria cardamomum*) is native to India. In Ayurvedic medicine, the seeds are considered to be one of the most effective digestive stimulants.**

Reducing the pressure

High blood pressure (hypertension) and angina, a condition where the heart's demand for blood is greater than that being supplied by the coronary arteries, may be managed by varying therapeutic strategies that involve the use of drugs with different mechanisms of action. Plants such as white hellebore and opium poppy contain chemicals that can lower blood pressure, and so have been developed as drugs themselves or have inspired the design of new pharmaceuticals used for heart conditions.

HELLEBORES AND THE HEART

PLANT:
Veratrum album L.
COMMON NAME(S):
white hellebore, European hellebore, white false hellebore
FAMILY:
trillium (Melanthiaceae)

ACTIVE COMPOUND(S):
NATURALLY OCCURRING:
steroidal alkaloids (protoveratrines A and B)
MEDICINAL USES:
MAIN: hypertension
OTHER: insecticide
PARTS USED:
root and rhizome

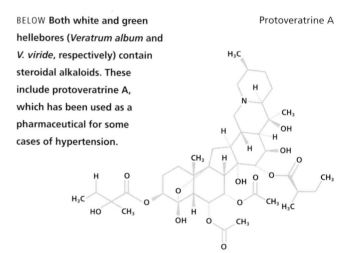

BELOW **Both white and green hellebores (*Veratrum album* and *V. viride*, respectively) contain steroidal alkaloids. These include protoveratrine A, which has been used as a pharmaceutical for some cases of hypertension.**

Protoveratrine A

ABOVE **Flowering white hellebore (*Veratrum album*) plants.**

White hellebore has a native range across parts of Europe, including Italy and Switzerland extending to Far East Russia, especially in mountainous regions. A closely related species in the same family is green hellebore (*Veratrum viride*), also known as the American hellebore, which is native to Alaska and the United States. Native Americans and early European settlers in North America utilized green hellebore for its medicinal properties and its use spread to England in the 1860s.

Both white and green hellebore contain steroidal alkaloids in their roots and rhizomes. These include protoveratrines A and B, which are common to both species. These compounds lower blood pressure by dilating blood vessels and slowing the heart rate. For this reason, they have been used in cases of severe hypertension – but only under controlled circumstances, because they can also produce toxic effects and so their use needs to be carefully monitored. It is for this reason that protoveratrines A and B have largely been replaced by other medicines for managing hypertension.

Decoctions of white hellebore were once used as insecticides. Although this use has also declined, a number of other, unrelated plants and their constituents have been studied or developed as antiparasitic agents against scabies and head lice that can infest humans (see page 176).

A POPPY ICON

The opium poppy is iconic in the history of medicine. Its therapeutic uses date back to ancient times, and opium from the poppy capsules has been linked with drug addiction and wars. Yet it has provided us with important drugs, critical for pain relief today (see page 140), and has been instrumental in developing other useful drugs for conditions ranging from coughs to Parkinson's disease (see page 154). In addition to the analgesic alkaloids morphine and codeine, another, chemically different alkaloid occurs in opium and has distinct biological properties. This isoquinoline alkaloid is papaverine. It lacks analgesic effects, but relaxes smooth muscles in the body to dilate blood vessels,

including the coronary arteries that supply blood to the heart. It also relaxes the heart muscle itself. Papaverine has therefore been prescribed for certain heart conditions.

Although not widely used now, papaverine was the inspiration for the design of the heart drug verapamil, which is classified as a calcium channel blocker because it interferes with the movement of calcium ions across heart cells and across the smooth muscle cells of blood vessels. This action reduces contractions of the heart muscle, influences the electrical impulses within the heart, and relaxes blood vessels to reduce blood pressure. Verapamil is therefore prescribed to manage certain forms of heart arrhythmias, hypertension and angina. The development of heart drugs from the opium poppy further reveals the diversity of its medicinal powers.

BELOW Green hellebore (*Veratrum viride*) is a herbaceous perennial that can grow to 2 m (6.5 ft) tall. The plant was traditionally used as an emetic, or externally as an analgesic.

Raiding the chemical messenger store

Indian snakeroot contains the indole alkaloids ajmaline, which has been used for cardiac arrhythmias (see page 40), and reserpine, which was once used to manage hypertension. Reserpine depletes stores of chemical messengers, especially noradrenaline (norepinephrine), in the nervous system, including in the heart and brain, resulting in reduced blood pressure, a slower heart rate and

depression of the central nervous system. In the same genus as Indian snakeroot is the milk bush (*Rauvolfia tetraphylla*), which is native to Mexico and tropical America. It is a source of the alkaloid deserpidine, which has similar properties to reserpine and has therefore also been used in the management of hypertension.

Thinning the blood

Clots may occur in our blood vessels for various reasons, and in some circumstances may result in a stroke or heart attack. Drugs used to help reduce the risk of blood clots are sometimes referred to as 'blood thinners'. They include warfarin, which was designed from chemicals occurring in mouldy melilot hay, and aspirin, which is based on salicylate chemicals from plants such as willow.

FROM BARN TO BEDSIDE

PLANT:
Melilotus officinalis (L.) Lam.

COMMON NAME(S):
melilot, yellow sweet clover,
king's clover

FAMILY:
legume (Fabaceae)

ACTIVE COMPOUND(S):

NATURALLY OCCURRING:
coumarins (dicoumarol)

SYNTHESIZED FROM NATURAL

LEAD: warfarin

MEDICINAL USES:

MAIN: anticoagulant

OTHER: circulatory disorders,
indigestion

PARTS USED:
aerial parts

RIGHT **Dicoumarol, formed from coumarin derivatives in mouldy melilot (*Melilotus officinalis*) hay by the action of microbes, was used to design the anticoagulant drug warfarin.**

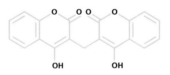

Dicoumarol

A mysterious disease in cattle in the 1920s is an unusual start for a blockbuster drug, but this is how the story of warfarin began. During the Great Depression, unexplained cases of livestock dying from internal bleeding occurred in the United States and Canada. In such times of hardship, fodder for livestock was in short supply, so the animals were grazed on mouldy melilot hay, which was investigated as the cause of the fatal bleeding condition that became known as sweet clover disease.

Outbreaks of this livestock disease continued for some years, before the specific cause was revealed in the 1940s. When sweet clover hay is infected with mould, the action of the mould microbes causes the coumarin chemical constituents in the plant to be converted into a new compound called dicoumarol, which has a powerful anticoagulant action, preventing blood from clotting. This discovery led to the idea of developing new dicoumarol derivatives for use as a rat poison. One of these new compounds was warfarin, which was marketed as a rodenticide in 1948. But the remarkable story of warfarin does not end here.

Warfarin acts by blocking the action of vitamin K. This modifies clotting factors in the blood and so inhibits coagulation, helping to prevent blood clots. For this reason, it was suggested that warfarin could be useful as a medicine to prevent blood clots, and so the first extensive studies in humans began in the 1950s. During this period, warfarin was famously given to United States President Dwight Eisenhower (1890–1969) after he had a heart attack. Today, warfarin is still the most widely prescribed oral anticoagulant drug, given to help prevent blood clots in certain heart conditions.

MEDICINE ON THE MOON

The modern pharmaceutical industry owes much to aspirin, a drug originally derived from salicylate chemicals that occur in plants such as willow, and meadowsweet (*Filipendula ulmaria*), which is in the rose family. The story of aspirin began in the eighteenth century, before it was discovered that salicylate chemicals could be useful to alleviate fevers, pain and inflammation (see page 142). Aspirin has even been taken to the moon by astronauts.

In the mid-twentieth century, aspirin's role as an analgesic had competition from new drugs that were being developed, such as the synthetic paracetamol (acetaminophen). But just as it seemed aspirin's place in medicine was waning, a new discovery catapulted it back into the limelight. Studies in the 1980s revealed that regular low doses of aspirin were linked with a reduced risk of certain heart disorders such as heart attacks. This is because aspirin blocks the action of the enzyme cyclo-oxygenase, in turn inhibiting the production of thromboxane A_2, a powerful inducer of platelet aggregation. When platelets aggregate in the blood, clots can form, so because aspirin inhibits platelet aggregation, it reduces the risk of clot formation. Today, aspirin is a key drug to help prevent blood clots in heart disease.

LEFT Melilot (*Melilotus officinalis*) is native to a region extending from Portugal to western Russia and south to Pakistan. This annual plant was introduced to North America in the early twentieth century as a cattle feed.

RIGHT Meadowsweet (*Filipendula ulmaria*) is a herbaceous perennial that can grow to 120 cm (47 in) tall and bears branched heads of creamy-white flowers that reward pollinators with pollen rather than nectar.

When garlic is crushed, sulfur-containing constituents such as alliin come into contact with the enzyme alliinase and are converted into different chemicals, including allicin. These chemicals may reduce cholesterol levels in the blood and might help prevent blood clotting. Consequently, there has been much interest in the use of garlic as a supplement, and in the diet, for any usefulness it may have in some forms of heart disease.

ABOVE Garlic (*Allium sativum*) bulbs and cloves.

Into circulation

Our circulatory system is a complex network of specialized blood vessels that carry blood around our bodies, transporting vital oxygen, nutrients and other substances. Blood circulation is maintained by the pumping action of the heart, but in some circumstances it may be impaired. Particular herbal medicines such as horse chestnut and butcher's broom have been the focus of studies into their potential usefulness in alleviating the symptoms of some circulatory problems.

FROM THE HORSE'S MOUTH

PLANT:
Aesculus hippocastanum L.
COMMON NAME(S):
horse chestnut
FAMILY:
soapberry (Sapindaceae)

ACTIVE COMPOUND(S):
NATURALLY OCCURRING:
triterpene saponins (aescin)
MEDICINAL USES:
MAIN: circulation
OTHER: rheumatism
PARTS USED:
seed

BELOW Horse chestnut (*Aesculus hippocastanum*) seeds contain aescin, which is a complex mixture of triterpene saponins, including aescin Ia.

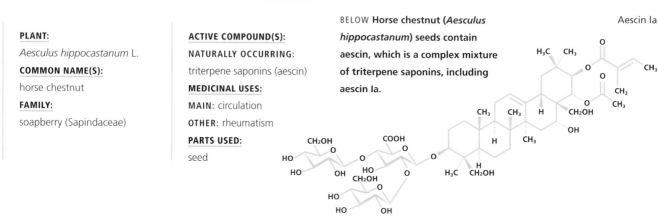

Aescin Ia

BELOW Horse chestnut (*Aesculus hippocastanum*) conkers (seeds), both loose and inside the prickly fruit.

Horse chestnut trees are native to an area stretching from the central Balkan Peninsula through to Turkey, but have been introduced to many other parts of the world, from Britain and Ireland to New Zealand. The spiny fruits encase glossy brown seeds that are also known as conkers. As well as being hung on a string and used in the children's game of the same name, they have been used traditionally for circulatory and other disorders (see box).

Laboratory studies have revealed that horse chestnut seed extracts have anti-inflammatory and antioedema properties, and they inhibit enzymes that are involved in the breakdown of the walls of capillaries, which are the minute blood vessels in our circulation. By helping to protect capillaries from damage caused by these enzymes, the extracts may therefore help to restore capillary function in circulatory disorders. The key active chemicals in the seeds are considered to be closely related triterpene saponins, a mixture known as aescin.

When the circulation in our legs is impaired in a condition known as chronic venous insufficiency, it may result in symptoms such as pain, oedema, itching and a feeling of 'heaviness' in the legs. Some clinical trials have concluded that horse chestnut seed extracts help alleviate some of these symptoms in people with the condition. There has also been interest in the use of seed extracts for alleviating sports injuries and bruising.

RIGHT **Stem tip of a female butcher's broom (*Ruscus aculeatus*) shrub. Its spine-tipped leaf-like cladodes bear a small female flower that is followed by a red berry.**

BUTCHER'S BROOM

Butcher's broom, which is also known by the more pleasant name of box holly, is native to parts of west-central Europe, including Britain and France. The plant's rhizomes were once taken as a decoction, and the berries and 'leaves' (cladodes) were applied as a poultice in the belief they would heal broken bones. The principal use of butcher's broom rhizomes in herbal medicine is to alleviate circulatory problems, especially those associated with the veins of the legs.

Some scientific studies suggest that extracts from the rhizomes might alleviate symptoms of chronic venous insufficiency and piles (haemorrhoids). The main active chemicals are considered to be steroidal saponins.

Seeds of time

It has been suggested that the name horse chestnut originates from the definitions 'large' or 'coarse' for the word 'horse'. Another explanation for this tree's common name is based on the doctrine of signatures, the belief that if a plant part resembles a part of the body, it could be useful in treating a disease of that origin. Part of the horse chestnut leaf (the cicatrix) is said to resemble a horse's hoof, so this could explain why ancient herbalists – including the English botanist John Gerard (*c.* 1545–1612) – claimed that the seeds of the tree could be a remedy for diseases in horses.

Another traditional belief was that if someone carried horse chestnut seeds or even wore them as a necklace, they would be protected from rheumatism and piles. Seed extracts were reputed to protect the skin from the sun, while seeds prepared as snuff were considered to cure headaches and catarrh. Horse chestnut seeds have long been used as a traditional remedy for leg vein disorders. Other parts of the plant have also been used traditionally, including the bark, which has been used for diarrhoea, piles and fevers, and externally for skin disorders. The leaves were once a folk remedy for coughs and rheumatic complaints.

CHAPTER 3

CALMING THE NERVES

The human nervous system is highly complex, controlling many systems in the body, including our circulation, breathing, movement, consciousness and mood. Dysfunction of our nervous system can result in a diverse range of diseases. Many plant compounds mediate effects on the nervous system, and some have been developed as important pharmaceuticals. This chapter explores examples of plant compounds that have been developed, or investigated, for their therapeutic value in disorders of the nervous system.

Common snowdrop (*Galanthus nivalis*)

NATURAL COMPOUNDS FOR THE NERVE NETWORK

Our nervous system is highly complex, consisting of different types of nerve cells (neurons) that have unique functions. Neurons receive, transmit and store information in a complicated network that controls many functions in our bodies, and disorders of the nervous system can affect our mood, consciousness, memory and movement. Many plant compounds can target our nervous system, including those developed as pharmaceutical drugs, and others used as traditional or complementary medicines.

THE NERVOUS SYSTEM: A COMPLEX CONCERT

Our nervous system is composed of billions of neurons working in concert to enable us to respond to the outside world and control internal bodily functions. It consists of the central nervous system (CNS), comprising the brain and spinal cord, which receives, processes and stores information; and the peripheral nervous system, which has two parts, sensory and motor. The sensory part picks up information on smell, sound, visual images, temperature and pain from our sensory organs and sends it to the CNS. The motor neurons of the peripheral nervous system receive messages from the CNS to control our muscles, including the voluntary muscles used for movement,

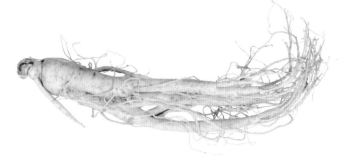

ABOVE **Ginseng (*Panax ginseng*; see page 64) root.**

and the involuntary muscles used to operate internal organs such as the heart and gut. We understand much about how our nervous system works, but because it is so complex and extensive, some details about its functions and potential drug targets are yet to be unravelled.

MOOD MARVELS

Many plants have been used traditionally to help calm the nerves, with some reputed to alleviate anxiety and stress. These include the rhizomes of kava (*Macropiper methysticum*, see page 56) in the pepper family (Piperaceae), which for centuries have been prepared as an intoxicating beverage for use in ceremonies in Oceania. Other plants are associated with alleviating insomnia, including valerian root (*Valeriana officinalis*; see page 58) in the honeysuckle family (Caprifoliaceae). St John's wort (*Hypericum perforatum*; see page 60), in the St John's wort family (Hypericaceae), has been used since the time of the ancient Greeks as a 'nerve tonic' and is now a popular herbal remedy for low mood. These and other plants used as complementary or alternative therapies to influence mood are often taken in the form of extracts containing mixtures of many

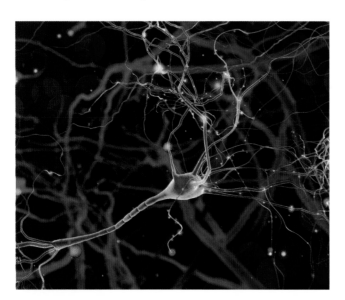

ABOVE **Our nervous system is complex and extensive, and there is still much to learn about how plants can be used to target nervous disorders.**

ABOVE Natural selection has favoured plants that produce chemicals that deter predation, some of which can influence nerve function. These have been harnessed both for medicinal applications and to develop insecticides, such as pyrethrin from the pyrethrum daisy (*Tanacetum coccineum*) in the daisy family (Asteraceae).

ABOVE The seeds of *Cola nitida* (see page 66), known as kola nuts, contain the stimulant alkaloid caffeine. Extracts from the nuts have been used as an ingredient for some soft drinks.

different compounds. Other plant compounds have stimulant effects, including caffeine from cola (*Cola nitida* and *Cola acuminata*) in the mallow family (Malvaceae), from coffee (*Coffea* spp.) in the coffee family (Rubiaceae) and from tea (*Camellia sinensis*) in the tea family (Theaceae) (see page 62).

MIND AND MEMORY

Some diseases cause neuron function to become impaired, so pharmaceutical interventions aim to correct this. One example is dementia, which is most commonly caused by Alzheimer's disease. Plants have had a key role in the discovery of drugs for dementia. Indeed, two of the five main drugs developed specifically to alleviate dementia symptoms are derived from plants. One is galantamine from members of the amaryllis family (Amaryllidaceae), including snowdrops (*Galanthus* spp.; see page 70). The other drug, rivastigmine, was designed from the plant compound physostigmine, found in the Calabar bean (*Physostigma venenosum*; see page 72) in the legume family (Fabaceae). Plant extracts containing many different compounds also have been tested for their usefulness in memory disorders, including dementia.

One example is the ancient ginkgo tree (*Ginkgo biloba*; see page 74) in the ginkgo family (Ginkgoaceae), whose genus name comes from the Japanese *ginkyo*, meaning 'silver apricot'.

NUMBING THE PAIN

Leaves of the Coca plant (*Erythroxylum coca*; see page 76) from the coca family (Erythroxylaceae) were a traditional medicine used to alleviate pain and, in the mid-nineteenth century, the alkaloid cocaine was isolated from them. Cocaine has anaesthetic properties and revolutionized surgical procedures; in addition, it inspired the development of some modern anaesthetics that are still used today. In a serendipitous discovery in the twentieth century, a mutant strain of barley (*Hordeum vulgare*; see page 77) in the grass family (Poaceae) was found to be the source of the alkaloid gramine. This plant chemical was used as a template to design other modern anaesthetics, further demonstrating how plants continue to surprise us as sources of important pharmaceuticals.

Life can be stressful

Throughout history, plants have been used for their reputed ability to alter our mood and help alleviate anxiety. They include the European herb valerian (see page 58), used especially for insomnia, as well as hops (*Humulus lupulus*), lemon balm (*Melissa officinalis*) and lavender (*Lavandula angustifolia*). Here, we look at kava, traditionally used in the South Pacific as a mind-altering beverage and also as a remedy for some nervous disorders.

RELAXING RHIZOMES

PLANT:
Macropiper methysticum (G.Forst.) Miq. (syn. *Piper methysticum* G.Forst.)
COMMON NAME(S):
kava, kava-kava
FAMILY:
pepper (Piperaceae)
ACTIVE COMPOUND(S):
NATURALLY OCCURRING:
kava lactones (kava pyrones),
including kawain (kavain), methysticin, yangonin
MEDICINAL USES:
MAIN: anxiety and stress
OTHER: other nervous conditions; infections of the genito-urinary tract
PARTS USED:
rhizomes

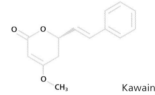

ABOVE **The chemical structure of the kava lactone kawain, one of the active components of kava (*Macropiper methysticum*) rhizomes.**

The perennial shrub kava has a narrow native range, from the Santa Cruz group in the Solomon Islands to Vanuatu, but was spread by early voyagers more widely across the islands of the Pacific, where it has a long history of use in ceremonies as an aromatic drink. Kava rhizomes have been popular in the Pacific islands for their psychoactive properties, but have also been used traditionally for a diverse range of conditions (see box). However, it is for conditions of nervousness – particularly states of anxiety and stress – that kava has been used as a herbal remedy elsewhere, particularly in North America, Europe and Australia.

In some laboratory studies, kava extracts and their constituent chemicals, the kava lactones, have been shown to alleviate anxiety and to have mild sedative effects. Scientific research has suggested that the kava lactones may modulate the action of some neurotransmitters in the brain to mediate these effects. Other studies suggest kava lactones have anticonvulsant and analgesic effects, that they may protect nerve cells, and that they have muscle-relaxant properties. The latter action might explain

LEFT **Kava (*Macropiper methysticum*) is used traditionally as an intoxicating and ceremonial beverage, and to alleviate anxiety.**

the traditional use of kava for restlessness. The results of some clinical trials have suggested that kava extracts may provide symptomatic relief of anxiety in humans, although more studies are needed to confirm these findings.

RESTRICTED RHIZOMES

Since the adoption of kava as a herbal remedy in Europe and elsewhere, its intake has been associated with a number of adverse effects, including rashes and jaundice. Importantly, the safety of kava was called into question following reports in 2000 and 2001 that its use could cause liver damage. By 2005, there had been almost 80 reports worldwide of liver toxicity associated with kava use, ranging from mild changes in liver function to severe damage resulting in the need for a liver transplant, or death. Consequently, kava preparations for internal use have been withdrawn or banned in some countries, including the UK, and safety warnings have been issued in others.

The zombie drink

For thousands of years, preparations made from kava have been consumed as an intoxicating beverage during ceremonies on islands of the South Pacific. Kava ceremonies were traditionally solemn events, restricted to men and, on some islands, to royalty and priests. To prepare the drink according to ritual, kava rhizomes were first chewed by young people, especially those with perfect teeth. The pulverized pulp was then mixed with water or coconut milk, and this was strained to obtain the kava beverage. Kava ceremonies are still performed today in some South Pacific island villages, during decision-making in the community or as a welcome for visitors.

The first Europeans to encounter kava were probably the botanist Daniel Solander (1733–1782) and artist Sydney Parkinson (1745–1771), who accompanied Captain James Cook (1728–1779) on his first voyage to the Pacific (1768–1771). In Australia, kava became a drug of abuse among Aboriginal communities following its introduction in the 1980s. Known as 'the killer kava', it was considered to turn people into zombies. In contrast, the kava plant has been highly regarded elsewhere for its reputed health benefits. In Hawaii, for example, kava was believed to restore strength and counteract weary muscles, it was used to alleviate chills and colds, and its leaves were prepared as a poultice to soothe headaches. Kava has also been used traditionally for asthma, urinary tract infections, rheumatism, fevers and syphilis, and even weight loss.

LEFT The use of kava as a mind-altering beverage is recorded in the species name *methysticum*, which means 'intoxicating', giving rise to the species' alternative common names of narcotic pepper and intoxicating pepper.

A TEAM EFFORT

PLANT:

Valeriana officinalis L.

COMMON NAME(S):

valerian

FAMILY:

honeysuckle (Caprifoliaceae, syn. Valerianaceae)

ACTIVE COMPOUND(S):

NATURALLY OCCURRING:

iridoids (valepotriates); volatile oil (sesquiterpenes, including valerenic acid)

MEDICINAL USES:

MAIN: insomnia, nervous conditions

OTHER: rheumatism, migraine, colic, cramp

PARTS USED:

rhizomes, roots

Valerian root has long been renowned for aiding sleep and calming the nerves. More recently, it has been extensively investigated for its relaxing and hypnotic effects, and some clinical trials suggest it may have a mild sedative action in humans. Valerian has also been of interest to alleviate anxiety and depression, but scientific research to date does not provide adequate evidence to support these uses.

Although valerian root has been the subject of numerous studies to determine its key biologically active chemicals, the sedative action cannot be attributed to a single chemical or class of chemicals alone. However, valerian is considered to act by the phenomenon of polyvalency. That is, it contains many different types of chemicals with different biological effects that work together to produce an overall therapeutic action. One class of chemicals in valerian is the iridoids, collectively known as the valepotriates. The valepotriates and their decomposition products (chemicals that valepotriates may break down to when they are prepared as extracts) show tranquillizing and muscle-relaxing actions. Other chemicals in the volatile oil from the root, including a substance called valerenic acid, enhance the action of the neurotransmitter gamma-aminobutyric acid (GABA) in the CNS, which decreases activity in the brain. Together, these chemicals act in different ways as a 'team' to produce an overall sedative effect.

SOOTHING POTENTIAL

Many other plants are reputed to have calming and sedative effects. However, the extent to which these have been investigated to ascertain whether there is any scientific basis for such uses varies considerably. The strobiles (papery female flower heads) from hops, which is in the cannabis family (Cannabaceae), have been used to flavour malt liquors and beer since the ninth century CE and have also long been used as a packing for pillows to aid sleep – King George III (1738–1820) is said to have used a hops pillow. Laboratory studies suggest hops extracts might have sedative and hypnotic properties, although more research is needed to establish whether they produce such effects in humans.

Lemon balm and lavender, both in the mint family (Lamiaceae), are also reputed to alleviate nervous conditions. Lemon balm has a distinct citrus odour and extracts from its leaves are said to have a sedative action. Lavender oil also has a characteristic fragrance and has been used either by applying it diluted to the

LEFT **Valerian (*Valeriana officinalis*), a herbaceous perennial up to 150 cm (60 in) tall, is widespread in Europe east to the Caucasus.**

skin or inhaling the volatile chemicals as a traditional remedy for agitation, anxiety, insomnia and other nervous disorders. More scientific research is needed to confirm whether lemon balm and lavender are useful in the nervous conditions they are reputed to help, although some experiments suggest lavender may be of some value for anxiety.

ABOVE **Flowering female hops (*Humulus lupulus*) plant.**

ABOVE **Lemon balm (*Melissa officinalis*) leaves.**

A smelly solution

The medicinal and aromatic value of valerian dates back to the eleventh century, when Anglo-Saxons believed the potent odour of the plant intoxicated cats and attracted rats. Indeed, some have suggested that valerian in the pockets of the Pied Piper, whose medieval story was retold in the nineteenth century by the brothers Grimm, was his charm against rats rather than his music. In contrast, the English herbalist John Gerard (*c.* 1545–1612) claimed that valerian could 'killeth mice'.

Although valerian is widely used in modern herbal medicine as a mild sedative, it once had many other uses. The seventeenth-century herbalist Nicholas Culpeper (1616–1654) recommended that valerian boiled in wine be taken by anyone 'Bitten or stung by any venomous creature'. Culpeper also claimed valerian could help to 'Expel wind in the belly', and that root and herb combined could be applied to alleviate pains in the head and rheumatism. Historically, valerian was known as all-heal in England. It was also called herb bennett, derived from *herba benedicta* or 'blessed herb', alluding to its reputation as a counter poison and its many medicinal uses, including as a remedy for bubonic plague.

ABOVE **An engraved portrait of the botanist Nicholas Culpeper.**

Depression and low mood

St John's wort, traditionally used to alleviate depression, has been extensively investigated to establish whether it has any effectiveness. Other plants such as the saffron crocus (*Crocus sativus*) in the iris family (Iridaceae) are also traditional remedies for nerve-related disorders, but have been tested less vigorously to understand whether there is any scientific basis for these medicinal applications.

ST JOHN'S WORT

PLANT:
Hypericum perforatum L.
COMMON NAME(S):
St John's wort
FAMILY:
St John's wort (Hypericaceae)
ACTIVE COMPOUND(S):
NATURALLY OCCURRING:
naphthodianthrones (hypericin); prenylated phloroglucinols (hyperforin)

MEDICINAL USES:
MAIN: depression
OTHER: externally for skin conditions, including wounds, burns, bruises and stings
PARTS USED:
aerial parts

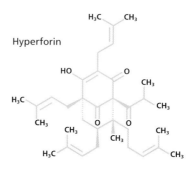

Hyperforin

LEFT **Hyperforin is a phloroglucinol derivative found in St John's wort (*Hypericum perforatum*) that is considered to contribute to the effects of extracts of the plant on the central nervous system.**

St John's wort is named because it typically flowers in June around the time of St John's Day (24 June). Its scientific species name, *perforatum*, refers to the translucent glands in the leaves, which resemble small perforations. The plant has been used since ancient times as a traditional remedy for nervous disorders, including anxiety, low mood and insomnia. There is much current interest in the use of St John's wort to alleviate symptoms of depression and it has been studied extensively.

One theory behind depression is that it is caused by decreased activity of some types of nerve cells in the brain. Current synthetic drugs aim to restore this activity by inhibiting enzymes called monoamine oxidases, which in turn inhibit the breakdown of certain neurotransmitters such as noradrenaline (norepinephrine). Another treatment strategy aims to prevent neurotransmitters such as serotonin from being taken up by nerve cells so that they remain at the junctions between nerve cells (synapses) for longer. Some synthetic antidepressant drugs act in this way. Both approaches increase the concentrations of neurotransmitters in the brain to restore or prolong their actions at nerve cells (see box).

LEFT **St John's wort (*Hypericum perforatum*) is a herbaceous perennial widely distributed from Europe to China, north to western Russia and south to Sudan.**

Laboratory studies show that St John's wort extracts and the plant's prenylated phloroglucinol constituents (especially hyperforin) inhibit the uptake of some neurotransmitters, so they appear to have a similar mechanism of action to some antidepressant drugs. In other studies, the naphthodianthrone constituent hypericin inhibited monoamine oxidase enzymes, while some flavonoid constituents could bind to particular nerve cell receptors, suggesting St John's wort may act by polyvalency (see page 58). Extracts of the plant have been tested in humans in clinical studies, which suggested they might be useful for some forms of depression. However, important interactions between St John's wort and a number of other medicines have been reported, in particular its effect of increasing the breakdown of certain drugs or increasing the action of others in the body, resulting in serious adverse consequences.

THE SUNNY SIDE OF SAFFRON

Saffron, the dried style and stigma of the saffron crocus flower, has been used in food and as a dye for centuries. It has also been used in some preparations for teething pain, while in traditional Chinese medicine it is considered to alleviate anxiety, fear, depression and confusion, and reputed to tranquillize the mind and create feelings of joy. Saffron extracts have been tested in small studies in humans, which suggested that they alleviate some symptoms of depression. Petal (tepal) extracts from the same species have also been tested in this way and showed similar effects. However, more studies are needed to evaluate whether extracts from either have any potential therapeutic benefits in conditions associated with low mood.

ABOVE **Saffron crocus (*Crocus sativus*) is native to Greece and a source of the expensive spice saffron.**

The secret of the snakeroot

Alkaloids from Indian snakeroot (*Rauvolfia serpentina*) in the dogbane family (Apocynaceae) have been investigated because of interest in this plant as a traditional Ayurvedic medicine for snakebites and nervous conditions. In 1952, the indole alkaloid reserpine was isolated from the root and was found to lower blood pressure. It was developed as a drug for this purpose (see page 47), but was then also discovered to have a depressive action. Subsequent studies ascertained that reserpine depletes the brain of neurotransmitters such as serotonin and noradrenaline. This revelation triggered research into the role of these neurotransmitters in depression and how they could be modified by drugs to alleviate symptoms. The discovery of reserpine and its mechanisms of action therefore revolutionized treatment strategies for depression and associated conditions.

ABOVE **Indian snakeroot (*Rauvolfia serpentina*) plant, with flowers and immature fruit.**

CALMING THE NERVES

61

EAST ASIA

The flora of China, the largest country in East Asia, comprises more than 31,000 species, or one-eighth of the world's plants, and is also the most diverse of any northern temperate region. Taiwan alone has nearly 4,100 plant species, which is a comparable number to the flora of the United Kingdom, yet it covers less then 7 per cent of the geographic area. Mongolia, to the north, is the second-largest country in the region, followed by the islands of Japan to the east and, lastly, by North and South Korea.

Ancient beginnings

The traditional medicine practised in China was the first East Asian system written down (see page 16) and has influenced the rest of the region through exchange of people and ideas between countries. Traditional Chinese medicine was at times resisted in Japan in favour of local medicines and practices, but in the twelfth century a system that included both traditional Chinese and Japanese medicines started to emerge. This led in time to the practice known today as Kampo.

For part of the twentieth century, traditional Chinese medicine was somewhat abandoned in China in favour of

FAMILIAR ORIENTAL PLANTS

Tea (*Camellia sinensis*), in the tea family, is probably the most well known plant from this region. It is widely used as a mild stimulant drink (see page 66), but also has medicinal applications for coughs, chest infections, osteoarthritis and certain types of wart (see page 173). Many other plants from East Asia's rich flora will also be familiar through their popularity in gardens in temperate regions of the world. These include trees such as magnolia (*Magnolia officinalis*) in the magnolia family (Magnoliaceae), shrubs like forsythia (*Forsythia* spp.) in the olive family (Oleaceae) and the herbaceous peonies (*Paeonia lactiflora*) in the peony family (Paeoniaceae). An intriguing feature of the flora is that around 65 genera of plants are found in the mountains of both East Asia and eastern North America, including ginseng (*Panax* spp.; see page 64) in the ivy family (Araliaceae), bugbane (*Actaea* spp.; see page 194) in the buttercup family, witch hazel (*Hamamelis* spp.; see page 205) in the witch hazel family (Hamamelidaceae) and magnolia vine (*Schisandra* spp.; see page 106) in the magnolia vine family (Schisandraceae). It is thought that these plant populations, now separated by thousands of kilometres, are remnants of a northern hemisphere flora dating back 34 million to 5 million years ago, arising from the connection of the continents by land bridges and similar climatic conditions.

ABOVE **Tea (*Camellia sinensis*) leaves being picked in the vicinity of Xishuangbanna, China.**

LEFT **Weighing out ingredients in a traditional Asian apothecary.**

BELOW **Dahurian bugbane (*Actaea dahurica*, syn. *Cimicifuga dahurica*) is one of the East Asian species of *Actaea* that have been used in unregistered 'black cohosh' preparations – intentionally or accidentally substituted for the correct species, *Actaea racemosa* (see page 194) – and may contribute to the serious health effects seen in women using these preparations.**

conventional medicine. Following the Communist Revolution of 1949, however, it was once more embraced as a way of reducing China's dependence on the West. Colleges teaching traditional Chinese medicine were set up and a more scientific approach was applied to the study of herbs and medical practices. During the Cultural Revolution of 1966–1976, doctors who practised conventional medicine were seen as intellectuals and replaced with herbal practitioners.

Conventional medicine also became popular in Japan and was predominant there in the late nineteenth and early twentieth centuries. Since the 1960s, however, Kampo has been used alongside conventional medicine, and many remedies are covered by the Japanese national health insurance programme.

The modern era

Today, traditional Chinese medicine is practised throughout China and is widely used, and even taught, alongside conventional medicine. It employs around 500 key plants as herbal drugs, including most of those mentioned in the box opposite. They are usually prescribed in combinations of up to ten, but sometimes many more. In addition, more than 4,500 plants are used in the folk medicines favoured by some members of the country's ethnic minorities. In contrast, Japanese Kampo uses fewer than 240 plants, many of which are also found in traditional Chinese medicine, along with some additional local species.

Traditional Chinese medicine has had a global influence. In the United Kingdom, for example, many towns have a traditional Chinese medicine practice, and the herbs used are being systematically added to the European Pharmacopoeia. Plants used in traditional Chinese medicine are also the source of some important pharmaceuticals described in this book, including sweet wormwood (*Artemisia annua*; see page 134) in the daisy family (Asteraceae), and Chinese ephedra (*Ephedra sinica*; see page 120) in the ephedra family (Ephedraceae). Many others are currently under investigation.

Just the tonic

Life can be physically and mentally challenging, so it is no surprise that a number of plants worldwide are used to help people get through the day. The roots of ginseng (*Panax ginseng*), a member of the ivy family (Araliaceae), have been studied extensively for their ability to increase physical endurance and improve mental capacity. Perhaps more widely used are plants containing the stimulant alkaloid caffeine (see page 66), which can counteract fatigue and help us feel more alert.

ADAPTING TO YOUR NEEDS

PLANT:
Panax ginseng C.A.Mey.
COMMON NAME(S):
ginseng
FAMILY:
ivy (Araliaceae)
ACTIVE COMPOUND(S):
NATURALLY OCCURRING:
saponins (ginsenosides)

MEDICINAL USES:
MAIN: tonic, adaptogen
OTHER: diuretic, stomachic
PARTS USED:
roots

Native to China, eastern Russia and Korea, ginseng is regarded as an adaptogen, a substance that can aid or normalize the body's response to stress. For this reason, it is often used as a herbal remedy to ward off illness rather than targeting a specific disease. Studies in humans have explored whether ginseng can improve physical performance, although its efficacy in this respect has not been established. Some clinical studies show ginseng may improve memory in humans with mild cognitive impairment or Alzheimer's disease, but more robust studies are needed to confirm this. There has also been interest in the use of ginseng to modulate the immune system, prevent cancer and manage diabetes. Again, the results of studies on its usefulness in these and other conditions have been varied. One explanation for the inconclusive effects of ginseng in humans is that different species of plants with the common name 'ginseng' are perhaps being used, and that these have different chemical properties and medicinal actions (see below).

LEFT **Ginseng (*Panax ginseng*) is a herbaceous perennial growing to 60 cm (24 in) tall. It bears a whorl of leaves and a central head of flowers that are followed by red fruit.**

The main chemical constituents of *Panax* species are the ginsenosides, which have been shown to have antiviral, nerve-protective and antioxidant properties. Other studies show that some ginsenosides can increase the force of contraction of the heart, while other ginsenosides have the opposite effect and reduce cardiac performance. These opposing effects might provide some explanation for why ginseng is described as an adaptogen.

A CASE OF MISTAKEN IDENTITY?

Many other species in the genus *Panax* are known by the common name 'ginseng', each of which has its own chemical ginsenoside 'fingerprint'. For example, American ginseng (*P. quinquefolius*) is indigenous to North America and is a reputed tonic and aphrodisiac. This species has been extensively traded and, combined with the decline in wild populations due to habitat loss, it is now endangered and its trade is regulated.

Another plant known as 'ginseng' or Siberian ginseng (*Eleutherococcus senticosus*), again in the ivy family, is also a reputed adaptogen and has been investigated for this action in humans. The roots of the plant contain eleutherosides, which differ from ginsenosides. Despite its 'ginseng' name, Indian ginseng (*Withania somnifera*) is in an entirely different plant family, the potato family (Solanaceae). It contains withanolides, steroidal saponins that have anti-inflammatory effects. Some withanolides inhibit the

enzyme acetylcholinesterase, an action associated with improved memory. The use of the common name 'ginseng' to describe different plant species has caused confusion when they are used or tested for their medicinal properties, so it is important that the correct species is identified.

ABOVE **The root of Indian ginseng (*Withania somnifera*) has been used traditionally as a tonic, especially against ageing, and has been tested for its effects on memory.**

Ancient panaceas

According to the ancient doctrine of signatures (see page 15), a plant that appears to resemble a part of the body can be useful in alleviating diseases that arise there. Ginseng roots have a human body-like appearance, which explains their reputation as a panacea. Indeed, the genus name *Panax* means 'all-cure' or 'all-heal'. Ginseng has been used for more than 5,000 years in traditional Chinese medicine as a tonic, and is believed to invigorate the body and prolong life. It has also been used as a remedy for frailty caused by long-term illness, and is said to tranquillize the mind and replenish wisdom.

Indian ginseng, also known as ashwagandha, is classed among the rejuvenating Rasayana tonics in Ayurvedic medicine (see page 45) and has been used for around 4,000 years. The third-century BCE Ayurvedic scholar Charaka said of this plant, 'One obtains longevity, regains youth, gets a sharp memory and intellect and freedom from diseases, gets a lustrous complexion and strength of a horse'.

LEFT **Miniature portrait of Charaka.**

FULL OF BEANS

PLANT:
Coffea arabica L.
COMMON NAME(S):
coffee, arabica coffee
FAMILY:
coffee (Rubiaceae)
ACTIVE COMPOUND(S):
NATURALLY OCCURRING:
purine alkaloids
(caffeine, theophylline)

MEDICINAL USES:
MAIN: stimulant
OTHER: fevers, flu, asthma
PARTS USED:
seeds

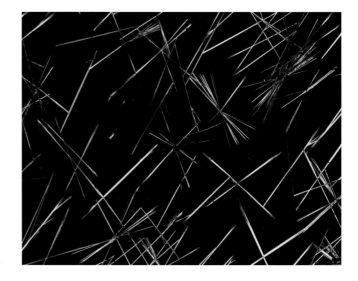

ABOVE **Chemical structure of the stimulant purine alkaloid caffeine.**

ABOVE **Polarised light micrograph of a section through caffeine crystals.**

What do coffee, tea and cola all have in common? In addition to being popular drinks, they are all made from plants that produce purine alkaloids, such as caffeine, which have stimulant properties. The caffeine in coffee beans (seeds) and tea leaves blocks the action of the neurotransmitter adenosine at its receptors in the brain. When adenosine binds to its receptors, it causes a biochemical reaction to promote drowsiness. By blocking the action of adenosine, caffeine counteracts this effect. This explains why drinking tea or coffee can stimulate our CNS, making us feel more awake and alert. Caffeine also inhibits an enzyme in the body called phosphodiesterase, which prevents the breakdown of a substance called cyclic adenosine monophosphate (cAMP), triggering chemical messengers to produce effects such as an increased heart rate.

Other plants also contain purine alkaloids such as caffeine and theophylline. These include species of cola, of which the dried seed leaves (cotyledons) – sometimes called kola nuts – are most famous for their original use in cola soft drinks (see also page 76). Another example is guaraná (*Paullinia cupana*) in the soapberry family (Sapindaceae), which was used traditionally as an anti-diarrhoeal, diuretic and aphrodisiac. In tropical regions of South America, where the plant is native, the seeds were crushed with water to produce a paste or powdered to prepare a stimulant tea. Guaraná is now used in herbal energy drinks and remedies that claim to combat fatigue. Paraguay tea, made from the leaves of maté (*Ilex paraguariensis*) in the holly family (Aquifoliaceae), is another caffeine-containing beverage used in South America and is reputed to alleviate headaches, joint pains and fatigue.

RIGHT **Arabica coffee (*Coffea arabica*) is becoming endangered in the wild, so scientists are investigating next-generation coffee hybrids that are better suited to changes in climate or that have improved resistance to pests and disease.**

ABOVE Guaraná (*Paullinia cupana*) fruiting shrub.

FROM MUG TO MEDICINE

There has been some interest in the use of coffee chemicals, including caffeine, to aid memory or even reduce the risk of some forms of dementia, but more research is needed to understand more fully any role they may have in cognitive disorders. As caffeine has stimulant properties, it is sometimes included in over-the-counter analgesic medicines used for headaches, including migraine.

In the 1880s, the London-based doctor Henry Hyde Salter (1823–1871), who had asthma, noticed that when he drank coffee his breathing improved. The ability of caffeine to dilate the airways was later found to be due to its inhibition of the enzyme phosphodiesterase. The related purine alkaloid theophylline was discovered to be an even more potent phosphodiesterase inhibitor, so is used as a medicine to alleviate asthma symptoms as it relaxes the airways and has a mild stimulant action on respiration. (For more about the use of plants for respiratory conditions, see Chapter 6.)

Threats, rescues and discoveries

The cultivation of coffee may have started in the sixth century CE in Yemen, where it was used for spiritual purposes. It is now an important commercial crop, and the second most valuable commodity worldwide after oil, with around 100 million people depending on the sale of it to support their livelihoods. The main species used as a beverage are arabica coffee (*Coffea arabica*), robusta coffee (*C. canephora*) and, to a lesser extent, liberica coffee (*C. liberica*). Arabica coffee in cultivation has a low genetic diversity, so cultivars may be vulnerable to climate change or attack by pests and diseases. Deforestation in some regions of Africa has led to arabica coffee becoming endangered in the wild, putting the future of our coffee at risk. To protect coffee and other useful plants, new approaches to storing seeds for the future are being developed.

There are 125 different species of coffee, which occur in Africa, islands of the Indian Ocean, parts of Asia and Australia. During recent botanical expeditions, scientists at the Royal Botanic Gardens, Kew discovered new members of the coffee family not previously known to science, including *Kupeantha* species in Cameroon and *Kindia gangan* in Guinea. These and many other species yet to be discovered need to be protected from threats if humans are to unlock their potential as future medicines.

LEFT *Kindia gangan*, first described in 2018, contains chemicals with biological activities of potential medicinal value.

Addiction therapy

Some chemical substances cause addiction in humans, with serious consequences for health. Examples derived from plants include nicotine from tobacco (*Nicotiana tabacum*) in the potato family, and cocaine from coca leaves (see page 76). Other plant chemicals that can mimic the actions or effects of more harmful addictive substances may be used as substitutes to help people during withdrawal.

CAUSE BECOMES CURE

PLANT:
Nicotiana tabacum L.
COMMON NAME(S):
tobacco plant
FAMILY:
potato (Solanaceae)
ACTIVE COMPOUND(S):
NATURALLY OCCURRING:
pyridine and pyrrolidine
alkaloids (nicotine)

MEDICINAL USES:
MAIN: smoking cessation aid
OTHER: insecticide
PARTS USED:
leaves

Nicotine

ABOVE **Chemical structure of the alkaloid nicotine, which is used as an aid for smoking cessation.**

The tobacco plant is native to Bolivia but has been introduced to other tropical, subtropical and temperate regions globally. Traditionally, tobacco was chewed, inhaled as snuff or smoked. It is still a recreational drug today, in the form of fermented leaves from the tobacco plant and related species, which are commonly smoked as cigarettes. Tobacco smoking is now known to have adverse effects on health, with some of its 3,000 chemicals associated with causing cancer. Tobacco is also addictive, which is attributed to the alkaloid constituent nicotine. Because nicotine stimulates cholinergic nicotinic receptors at the junction between nerve and muscle cells, and in the CNS, it produces complex effects on the body, including in the brain, the respiratory system and the heart. The addictive nature of nicotine makes it hard for people to stop smoking, but providing an alternative source of the pure compound – such as in the form of gums, lozenges, nasal sprays or patches – can help to reduce the withdrawal symptoms.

Other plants contain chemicals that can stimulate nicotinic receptors, so have also been the focus of work to develop drugs to aid smoking cessation. One example is the alkaloid cytisine, which occurs in some plants in the legume family, including laburnum (*Laburnum anagyroides*), a tree also known as golden rain. Cytisine is toxic, but its actions on the CNS resemble those of nicotine and its chemical structure has inspired the design of a new drug. Called varenicline, this has a similar action to nicotine and so has been developed to aid smoking cessation. The alkaloid lobeline from Indian tobacco (*Lobelia inflata*) in the bellflower family (Campanulaceae) also has similar effects to nicotine and has been explored for its therapeutic potential as a smoking deterrent. However, its usefulness for this purpose has not yet been confirmed.

HALLUCINOGENIC ROOTS

The roots of the iboga shrub (*Tabernanthe iboga*) in the dogbane family were used traditionally as a stimulant and as a hallucinogen in parts of tropical Central Africa, where it is native (see page 108). In Gabon, the roots and stems have been used as a traditional remedy for diabetes. The roots contain the indole alkaloid ibogaine, which modulates the action of some neurotransmitters in the brain, including dopamine. Ibogaine has been investigated because of its claimed usefulness in combating addiction to alcohol and some drugs such as cocaine. However, when tested on people with drug dependence, ibogaine has caused hallucinations and serious adverse effects on the nervous system and heart, and some fatalities have occurred as a result of its use.

ABOVE Indian tobacco (*Lobelia inflata*) is an annual or biennial herbaceous plant native to eastern North America and can grow to 100 cm (40 in) tall. It was traditionally used to aid breathing in conditions such as asthma, and alternative common names such as pukeweed indicate the plant's emetic effects.

LEFT Tobacco (*Nicotiana tabacum*) flowering plants can grow to 2 m (6.5 ft) tall. The leaves are used in cigarettes, which have been linked to adverse effects on health.

The buzz about nicotine

There are records of tobacco being used as an insecticide to combat aphids as far back as the 1760s. Species of the tobacco genus contain insecticidal alkaloids, including nicotine, nornicotine and anabasine. Insects have nicotinic receptors in their nervous system, so by stimulating these receptors, nicotine and related compounds have toxic effects on their nerve function. For this reason, nicotine-related insecticides, sometimes known as neonicotinoids, have been developed commercially since the 1990s.

While neonicotinoids have been used to protect crops from damage caused by insect pests, there has been some concern that they may be contributing to the current decline in bee populations. Without pollinators such as bees, the security of global food crops and the biodiversity of our planet will be at risk. As a result, there are some restrictions on the use of particular neonicotinoids, and research continues to unravel their effects on bees.

Dementia and memory

Two of the five main drugs developed specifically to alleviate dementia symptoms, particularly in Alzheimer's disease, are derived from plants. One is galantamine, which occurs in members of the amaryllis family, including snowdrops, daffodils (*Narcissus* spp.) and snowflakes (*Leucojum* spp.). The other, rivastigmine, is based on a chemical found in the Calabar bean (see page 72). Other plants such as ginkgo (see page 74) are used as herbal extracts to alleviate memory problems.

LEFT **The alkaloid galantamine inhibits the enzyme acetylcholinesterase and is a licensed medicine used to treat symptoms of dementia in Alzheimer's disease.**

Galantamine

BELOW **Galantamine was first isolated from Woronow's snowdrop (*Galanthus woronowii*), which is native to Turkey, Russia and Georgia.**

A JOURNEY OF DISCOVERY

PLANT:
Galanthus woronowii
Losinsk.

COMMON NAME(S):
Woronow's snowdrop

FAMILY:
amaryllis (Amaryllidaceae)

ACTIVE COMPOUND(S):
NATURALLY OCCURRING:
isoquinoline-derived alkaloid
(galantamine)

MEDICINAL USES:
MAIN: dementia in
Alzheimer's disease
OTHER: myasthenia gravis,
poliomyelitis

PARTS USED:
bulbs

From early associations with bad luck and death, and few records of traditional medicinal uses, snowdrops have emerged as triumphant in their role in drug discovery for dementia. Snowdrops have a mainly Mediterranean distribution, stretching east into Eurasia, with several species growing in southwest Russia, Georgia, Turkey and Bulgaria. In the mid-twentieth century, it was believed that rubbing snowdrops on the forehead eased nerve pain, and there are records of snowdrop decoctions being given to children with poliomyelitis symptoms in the Caucasus mountains. These were the first clues that compounds in snowdrops might be useful for nerve disorders.

In the 1950s, the alkaloid galantamine was isolated from Woronow's snowdrop (*Galanthus woronowii*) and was found to inhibit the enzyme acetylcholinesterase. This provided the first scientific basis for its use in nerve-related disorders. Later that decade, galantamine was discovered in other members of the amaryllis family, including the common snowdrop (*Galanthus nivalis*), daffodils and snowflakes, occurring in the leaves of these plants as well as their bulbs.

Galantamine inhibits acetylcholinesterase, so prolongs the action of the neurotransmitter acetylcholine between nerve and muscle cells, and between (cholinergic) nerve cells, enhancing or restoring their function. From the 1950s in parts of eastern Europe, galantamine was therefore used to treat poliomyelitis and myasthenia gravis, and after surgery to reverse the effects of some muscle-relaxant drugs. However, as new drugs with advantages over galantamine became available, its use gradually declined – until a new discovery that revolutionized current treatment strategies for dementia.

A LATE BLOOMER

In the 1970s, pioneering research revealed that in the brains of those with Alzheimer's disease, deficits in cholinergic neurons occur – notably in the hippocampus, which is important for learning and memory. This led to the hypothesis that drugs prolonging or mimicking the action of acetylcholine might restore cholinergic neuron function and so reduce Alzheimer's symptoms. Galantamine not only inhibits acetylcholinesterase, but it also mimics acetylcholine and stimulates the relevant (nicotinic) receptors on nerve cells. With this dual action to improve cholinergic signals in the brain, galantamine was tested in clinical trials in Alzheimer's patients, where it was found to improve symptoms.

As the trade of snowdrops across international borders is regulated to ensure wild populations are not endangered, an alternative source of galantamine was needed to meet demand for medicinal development. Although galantamine can be chemically synthesized in laboratory conditions, it can also be commercially extracted from cultivated daffodils, providing a

sustainable source. So, just as it seemed the medicinal value of galantamine was waning, these new revelations resulted in the first licences being granted in the UK, Ireland and elsewhere in 2000 to allow the drug to be used to treat dementia symptoms in Alzheimer's disease.

ABOVE **Summer snowflake or Loddon lily (*Leucojum aestivum*) is a riverside plant native to Europe and eastern Iran.**

BELOW **The daffodil cultivar *Narcissus pseudonarcissus* 'Carlton' has been used as a source of galantamine for pharmaceutical applications.**

A legend of luck

According to legend, it was considered unlucky to bring snowdrops indoors, and doing so was believed to cause cow's milk to become watery and to ruin butter making, or even lead to the death of someone in the house. Snowdrops were also known as death's flower because the flowers were considered to resemble a shroud. In contrast, it was also said that bringing snowdrops into a house at Candlemas would drive away evil, and that anyone who found a snowdrop before the first of January would have a lucky year.

One of the oldest records hinting at the medicinal properties of snowdrops is the *Odyssey* by the ancient Greek poet Homer. Here, 'moly' was used by Odysseus as an antidote against Circe's poisonous potion. The 'moly' had 'A flower as white as milk' and is believed to have been the snowdrop. Indeed, the scientific name *Galanthus* is derived from the Greek word *gala*, meaning 'white'.

POISONS AND ANTIDOTES

PLANT:
Physostigma venenosum Balf.

COMMON NAME(S):
Calabar bean

FAMILY:
legume (Fabaceae)

ACTIVE COMPOUND(S):

NATURALLY OCCURRING:
indole alkaloid
(physostigmine)

SYNTHESIZED FROM

NATURAL LEAD: rivastigmine

MEDICINAL USES:

MAIN: dementia in
Alzheimer's and Parkinson's
disease

OTHER: ordeal poison

PARTS USED:
seeds

RIGHT **The alkaloid physostigmine has been used in the design of drugs such as rivastigmine, used for some nervous system disorders, including dementia.**

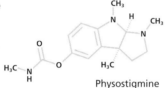

Physostigmine

ABOVE **The highly toxic Calabar beans (*Physostigma venenosum*) were used traditionally as an ordeal poison.**

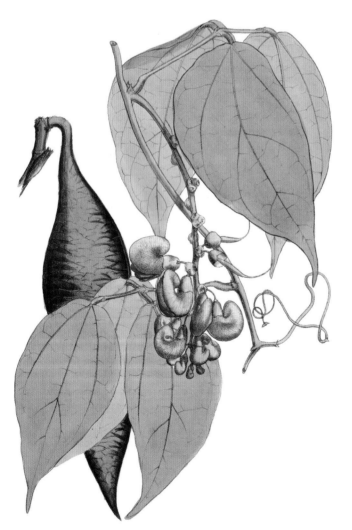

ABOVE **Calabar bean (*Physostigma venenosum*) is a scrambling or climbing forest plant with stems that can reach 15 m (50 ft) in length. Pollinated flowers develop into pods containing two or three seeds.**

Plants and their constituents that have potent effects on the nervous system can be poisonous, but this action also provides clues in the discovery of medicines that can target nerve disorders. One such plant is a climber native to tropical West and western Central Africa that produces the highly poisonous Calabar beans.

In the 1840s, Reverend Zerub Baillie (1831–1865), a Scottish missionary in what is now southern Nigeria, noticed that people who ingested Calabar beans experienced toxic effects, including excessive salivation, a flushed face and protruding eyes. British pharmacologists investigated the Calabar bean and found its actions in the body were the opposite to those of a different alkaloid called atropine, which occurs in poisonous plants from the potato family (see page 33), including deadly nightshade, also known as belladonna (*Atropa bella-donna*), and mandrake (*Mandragora officinarum*; see page 180). Atropine blocks the action of the neurotransmitter acetylcholine and produces such effects as dilating the pupils, reducing saliva and other secretions, and increasing heart rate. It was therefore concluded that extracts from the Calabar bean could act by enhancing the action of acetylcholine, and that it might be useful to treat poisoning caused by ingestion of atropine-containing plants. Indeed, by the 1860s a Calabar bean extract was used to treat a case of atropine poisoning in Prague. Calabar beans contain indole alkaloids, including physostigmine (also known as eserine), which was first isolated in 1864. It was not until the 1920s, however, that physostigmine was confirmed to inhibit acetylcholinesterase, thereby prolonging the action of acetylcholine. This explained the toxicity of Calabar beans, and provided the basis for their future medicinal value.

Calabar beans were used traditionally in Africa, particularly in the region of Old Calabar in Nigeria, for ritual deaths associated with the funeral of a chief and as an ordeal poison. Suspects accused of a crime were given a preparation of the beans, either crushed or in the form of a deadly potion. Any who survived were considered innocent, but those who died were believed to be guilty and paid the ultimate penalty. This logic may have some scientific basis, as nervous sipping by guilty suspects would enable greater absorption of the toxic alkaloids.

The poisonous effects of Calabar beans have also played a more recent role in crimes. In 2001, the headless and limbless body of a boy was found in the River Thames near Tower Bridge in London. The police at Scotland Yard named the victim 'Adam'. When the contents of his lower intestine were sent for identification at the Royal Botanic Gardens, Kew, fragments of Calabar beans were identified there by a botanist. The police believe these toxic beans may have been used to subdue the victim.

DISCOVERING AND DESIGNING DRUGS

Physostigmine was used as a template to design new acetylcholinesterase-inhibitor drugs such as neostigmine, used for symptoms of muscle weakness in myasthenia gravis. From the 1980s, physostigmine was tested for its usefulness in Alzheimer's disease, but benefits on memory were inconclusive and side effects were common. Another drug with the same mechanism of action was designed, again based on physostigmine. Called rivastigmine, it improved symptoms in dementia, so became available as a drug for this use, both as an oral medicine and in the form of patches applied to the skin.

Many other plants contain chemicals, especially alkaloids, that also inhibit acetylcholinesterase and so have become the focus of interest for their potential therapeutic value. The toothed clubmoss (*Huperzia serrata*, syn. *Lycopodium serratum*) in the clubmoss family (Lycopodiaceae) has been used in traditional Chinese medicine for conditions associated with memory loss and contains a potent acetylcholinesterase-inhibiting alkaloid called huperzine A. It is included in some remedies for alleviating impaired cognitive functions. Huperzine A has been tested in people with Alzheimer's disease with some promising effects, although more studies are needed to evaluate it further.

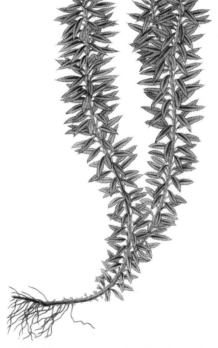

TOP AND RIGHT **The toothed clubmoss (*Huperzia serrata*) has a wide distribution, being native to parts of Asia, Malesia, Oceania and Central America. Its branched stems can reach 15–30 cm (6–12 in) in height.**

A LIVING FOSSIL

PLANT:

Ginkgo biloba L.

COMMON NAME(S):

ginkgo, maidenhair tree

FAMILY:

ginkgo (Ginkgoaceae)

ACTIVE COMPOUND(S):

NATURALLY OCCURRING:

diterpenes (ginkgolides);
sesquiterpenes (bilobalide);

flavonoids (including
biflavones and flavonols)

MEDICINAL USES:

MAIN: circulatory and
cognitive disorders

OTHER: asthma,
cardiovascular disorders

PARTS USED:

leaves

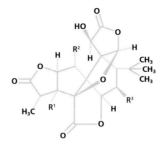

	R¹	R²	R³
Ginkgolide A	OH	H	H
Ginkgolide B	OH	OH	H
Ginkgolide C	OH	OH	OH

ABOVE **The ginkgolides have cage-like chemical structures and antagonize the action of platelet-activating factor, which is thought to reduce the risk of blood clots.**

ABOVE **Leaves of ginkgo or maidenhair tree (*Ginkgo biloba*) with their autumnal yellow coloration. Ginkgo is native to south central and southeast China, and is considered endangered in the wild, but it is widely cultivated including as an ornamental.**

Some plant extracts have been investigated for their effects on memory, and those showing promise have been tested for their usefulness in dementia. Often, their activities cannot be attributed to a single chemical; instead, a mixture of chemicals is believed to act in different ways to produce an overall effect in the body. Ginkgo is an example of a plant that acts by this phenomenon, known as polyvalency. Other plants have been tested for their effects on memory because of their uses in traditional medicine, including those in the mint family (see box).

Ginkgo is known as a living fossil as it is the only surviving member of an ancient group of trees that date back to the time when dinosaurs roamed the Earth. This Jurassic giant of trees can reach up to 25 m (80 ft) in height. In autumn, the leaves become a strikingly bright yellow colour due to the loss of the green chlorophyll, which exposes the high levels of yellow carotenoids and fluorescent-like chemicals that act like optical brighteners – hence ginkgo's alternative common name, maidenhair tree.

Ginkgo seeds and leaves are used traditionally in China for respiratory complaints. It was not until the 1960s that the leaves became popular in Europe for circulatory disorders, such as leg pain caused by insufficient blood flow (intermittent claudication) and ringing in the ears (tinnitus). Ginkgo leaf extracts were tested in humans and gave some encouraging effects on memory, so this stimulated interest in testing extracts in dementia patients. Early studies concluded that there was promising evidence that ginkgo extracts could improve cognitive functions in people with Alzheimer's disease, but other studies found that evidence for any benefits was inconsistent and unreliable. Therefore, more studies are needed to confirm the actions of ginkgo in dementia.

Ginkgo contains unusual diterpene chemicals called ginkgolides, with cage-like structures that antagonize the action of a substance in the body called platelet-activating factor (PAF). By antagonizing PAF, the ginkgolides inhibit the activation of platelets in the blood, which may reduce the risk of blood clots. This might improve blood flow in the brain, which is considered to contribute to the actions of ginkgo on memory, although there has been concern that ginkgo extracts may interact with some drugs that reduce blood clotting. The ginkgolides, along with other ginkgo constituents – including the sesquiterpene bilobalide and flavonoids – have shown antioxidant and anti-inflammatory effects. Some studies suggest they may also help protect nerve cells and modulate some neurotransmitters in the brain such as acetylcholine.

THE LESSER PERIWINKLE

The lesser periwinkle (*Vinca minor*) in the dogbane family has a native range extending from Europe to the Caucasus. The leaves contains an indole alkaloid called vincamine, which increases blood circulation in the brain and so has been of interest for use in some cognitive disorders. A synthetic derivative of vincamine is vinpocetine, which studies suggest may also increase blood flow to the brain and improve memory. It has therefore been suggested for use in vascular dementia, but further studies are needed to assess whether it has any therapeutic value.

ABOVE Lesser periwinkle (*Vinca minor*) is a shrub with trailing stems that can grow to 30–60 cm (12–24 in) in length and shorter erect stems that bear blue-purple to white flowers.

The marvellous mint family

Long before Aloysius Alzheimer (1864–1915) described Alzheimer's disease in 1906, old English herbals indicated that some plants in the mint family had a reputation for improving memory. In his sixteenth-century herbal, for example, John Gerard claimed that sage (*Salvia officinalis*) was 'Singularly good for the head and brain and quickenethe the nerves and memory', while in 1756, English botanist John Hill (c. 1714–1775) said that the herb 'Will preserve faculty and memory more valuable to the rational mind than life itself'. A few years earlier, Hill declared that lemon balm (*Melissa officinalis*; see page 59) was 'Good for disorders of the head'. Nicholas Culpeper reported that rosemary (*Salvia rosmarinus*, syn. *Rosmarinus officinalis*) helps 'The head and brain' and 'Dullness of the mind'. Modern studies in humans have suggested extracts from these plants might improve memory, while peppermint (*Mentha × piperita*; see page 90) was also found to improve

subjects' ability to perform some cognitive tasks. These studies merit more research on the effects of the marvellous mint family on memory.

ABOVE Sage (*Salvia officinalis*) is a woody herbaceous plant native to southwest Germany and southern Europe.

Numbing the pain

The coca plant is a source of cocaine, which was the basis for the development of modern anaesthetics that are used today. Some other plants produce chemicals that have enabled new anaesthetic drugs to be designed, while others such as cloves (the dried flower buds from *Syzygium aromaticum*) in the myrtle family (Myrtaceae) have been used in the form of extracts or oils for the relief of pain because of the anaesthetic-like action that their chemical constituents induce.

DREAMS, DRINKS AND DRUGS

RIGHT **Chemical structure of the alkaloid cocaine from the coca plant (*Erythroxylum coca*), which was used to design the anaesthetic drug procaine.**

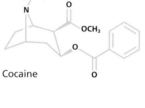

Cocaine

PLANT:	**SYNTHESIZED FROM NATURAL**
Erythroxylum coca Lam.	**LEAD:** procaine
COMMON NAME(S):	**MEDICINAL USES:**
coca	**MAIN:** anaesthesia
FAMILY:	**OTHER:** pain, nausea and
coca (Erythroxylaceae)	vomiting
ACTIVE COMPOUND(S):	**PARTS USED:**
NATURALLY OCCURRING:	leaves
tropane alkaloid (cocaine)	

ABOVE **Light micrograph of crystals of cocaine, a tropane alkaloid obtained from the leaves of the coca plant (*Erythroxylum coca*).**

ABOVE **The coca plant (*Erythroxylum coca*) is a shrub or small tree native to western South America, from Colombia to Bolivia.**

The story behind the discovery of cocaine, one of the earliest anaesthetics, begins more than 2,000 years ago, when the coca plant in which it occurs was reputed to provide pain relief. The plant has also been used in South America as a masticatory with a stimulant action. In 1860, the tropane alkaloid cocaine was isolated from coca leaves. Later that century, coca leaves were reported to improve digestion and to increase mental alertness and physical vigour, and cocaine was being investigated as a treatment for addiction to the drug morphine, which occurs in the opium poppy (*Papaver somniferum*; see page 140). Perhaps inspired by these reports, Sigmund Freud (1856–1939), a doctor from Vienna who later became a well-known psychiatrist, tried to help cure a friend of morphine addiction by giving him cocaine. Unfortunately, Freud's friend became addicted to cocaine and he later died. Freud was still convinced that cocaine could be useful, so he gave a cocaine preparation to a colleague, aiming to cure his intestinal pains. His friend declared the cocaine liquid numbed his lips and tongue, inspiring another colleague, the Austrian ophthalmologist Karl Koller (1857–1944), to develop it as an anaesthetic for use in eye surgery. Koller became famous for his discovery and was known as 'Coca Koller', reputedly much to the annoyance of Freud and shattering the latter's dreams of fame at that time.

Meanwhile, US pharmacist John Stith Pemberton (1831–1888) developed a coca leaf and caffeine drink, Wine Coca, in 1885. It tasted unpleasant, so Pemberton added kola nut extract to it and called it Coca-Cola; it was launched in 1886. Following a change of ownership in 1903, the cocaine was removed and sold to the pharmaceutical industry. This is the origin of the cola soft drinks used today. Cocaine revolutionized surgical procedures when its anaesthetic properties were confirmed, but it was realized the drug could cause addiction and toxicity, so its use is now restricted. The chemical structure of cocaine has been used to design modern local anaesthetics such as procaine.

MUTANT BARLEY

While we are most familiar with barley (*Hordeum vulgare*) as an important food crop, a serendipitous discovery in the 1930s resulted in a new anaesthetic. Investigations on a chlorophyll-deficient mutant strain of barley revealed that it contained an alkaloid called gramine. A similar chemical called *iso*-gramine was synthesized, and when this was tested, it was found to numb the tongue. A series of related chemicals were also synthesized, leading to the discovery of a new anaesthetic drug, lidocaine, which is still used in surgical procedures today, including in dental practice.

The electric daisy

Electric daisy, eyeball plant and toothache plant are all common names for *Acmella oleracea*, in the daisy family, whose flower heads were chewed as a traditional remedy for toothache in the Amazon region. The plant contains a chemical called spilanthol, which has mild anaesthetic-like properties. For this reason, extracts of the flower heads and spilanthol have been of interest in the development of a remedy for toothache, rubbed on the gums in the form of a gel, and have been used in skin preparations to relieve joint pain.

Cloves have been used traditionally to alleviate colic and flatulence, and as a toothache remedy. Cloves contain an essential oil composed of fragrant volatile chemicals, including eugenol, which has a local anaesthetic action. Clove oil has been included in some preparations for musculoskeletal and joint disorders, and is sometimes still used as a remedy for toothache.

ABOVE Electric daisy is a recent common name for the species *Acmella oleracea*, whose flower heads cause a tingling and numbing sensation in the mouth if chewed.

CHAPTER 4

IN SICKNESS AND IN HEALTH

A well-functioning digestive system is important for overall health but is often disrupted by disease and modern living. Here, we look at the plants and plant compounds that can help to restore balance when things have gone wrong, combatting sickness, ulcers, diarrhoea, constipation and attack by intestinal parasites.

Henbane (*Hyoscyamus niger*)

NATURAL COMPOUNDS FOR THE DIGESTIVE SYSTEM

Our body relies on the absorption of nutrients released by the digestion of food for the fuel and materials it needs. Yet the digestive system, and the body as a whole, is constantly under attack from harmful compounds and organisms we ingest, intentionally or otherwise, and is also affected by our choice of food and lifestyle. Some plant compounds have effects on our digestive system that can be harnessed to counteract diseases or unwanted symptoms.

ABOVE **Russian henbane (*Scopolia carniolica*; see page 86) is native to Italy and northeast to the Baltic states and Ukraine, and was named after the Austrian-Italian botanist and physician Giovanni Antonio Scopoli (1723–1788). It is a source of hyoscine and is used medicinally in Japan and Korea.**

SICK TO THE STOMACH

Nausea and vomiting are among the earliest and commonest symptoms of eating a poisonous plant. However, a few plants can calm a nauseous stomach, the principal among these being ginger (*Zingiber officinale*; see page 82), in the ginger family (Zingiberaceae). It is probably no coincidence that ginger is widely used as an addition to food and is a popular herbal tea, as well as being taken medicinally.

Particular causes of sickness may be treated with other plant compounds. Chemotherapy is an aggressive treatment for cancer that can have nausea and vomiting as a side effect. Synthetic derivatives of the psychoactive *delta*-9-tetrahydrocannabinol (THC) from cannabis (*Cannabis sativa*; see page 83) in the cannabis family (Cannabaceae) have been developed to reduce these symptoms. Motion sickness, including travel sickness, can be alleviated by small doses of the toxin hyoscine (scopolamine) and its derivatives. Hyoscine occurs in several members of the potato family (Solanaceae), including black henbane (*Hyoscyamus niger*; see page 86).

ULCERS AND WORMS

The roots of common liquorice (*Glycyrrhiza glabra*) in the legume family (Fabaceae) might principally be used in confectionary in the west, but in Asian countries they are often used as an ingredient in herbal preparations for spleen and stomach deficiencies that have reduced the body's levels of *qi*, or vital force. Liquorice may have some benefit in the healing of gastric ulcers, but at treatment doses it can have side effects. Similar side effects are caused by a semi-synthetic derivative of the liquorice compound glycyrrhizic acid, called carbenoxolone sodium (see page 88). Plant-based treatments for intestinal worms – including arecoline from the seeds of the betel palm (*Areca catechu*; see page 98) in the palm family (Arecaceae) – have also largely been replaced by safer synthetic alternatives.

RIGHT **Plants that were traditionally used to kill intestinal worms are called anthelmintics, and this is often reflected in their scientific names, such as the prickly ox tongue (*Helminthotheca echioides*, syn. *Helminthia echioides*) in the daisy family (Asteraceae).**

TOO FAST OR TOO SLOW

Peppermint (*Mentha × piperita*; see page 90) in the mint family (Lamiaceae) has long been used to flavour food and confectionary, as a herbal tea and as a medicinal plant, but peppermint oil is now used to relieve symptoms of irritable bowel syndrome (IBS). Constipation and diarrhoea, some of the symptoms of IBS, can have many other causes. To treat constipation, plants make significant contributions as bulk-producing laxatives, such as seeds of ispaghula (*Plantago ovata*), in the speedwell family (Plantaginaceae), and stimulant laxatives (see page 92). The stimulant laxatives exploit a common effect of some toxic plants, that is to cause diarrhoea, but used at a controlled dose. Hydroxyanthracene compounds are found in plants from several families, including the legume, buckthorn (Rhamnaceae), knotweed (Polygonaceae) and asphodel (Asphodelaceae) families. As with many plant chemicals that have medicinal applications, these compounds have a defensive role within the plant, in this instance being antimicrobial.

Diarrhoea and dysentery are a particularly heavy burden on the health of people in developing countries who don't have access to clean drinking water or sanitation. Although there are numerous traditional remedies, few have become commercialized. The most widely used include morphine from the opium poppy (*Papaver somniferum*; see page 97) in the poppy family (Papaveraceae) and its synthetic derivatives. Other remedies include beechwood creosote (see page 96) and crofelemer from dragon's blood (*Croton lechleri*; see page 94) in the spurge family (Euphorbiaceae), which shouldn't be confused with other plants commonly called dragon's blood, such as *Dracaena draco* in the asparagus family (Asparagaceae) and *Calamus draco* (syn. *Daemonorops draco*) in the palm family.

TOP AND ABOVE **The dried ripe seeds of ispaghula (*Plantago ovata*) and some other species of the plantain genus (*Plantago* spp.) are used as a bulk-forming laxative.**

Hard to stomach

Nausea and vomiting are the body's natural responses to eating something poisonous, infected or to excess. They can also be triggered in other ways, such as during pregnancy, abnormal motion and medical interventions. Traditional systems of medicine include many plants to soothe the stomach, such as ginger. And we can thank cannabis for a synthetic antiemetic used during chemotherapy, and members of the potato family for medicines to relieve motion sickness (see page 86).

WARMING GINGER

PLANT:
Zingiber officinale Roscoe
COMMON NAME(S):
ginger
FAMILY:
ginger (Zingiberaceae)

ACTIVE COMPOUND(S):
NATURALLY OCCURRING:
phenolic ketones (gingerols, shogaols) in oleoresin
MEDICINAL USES:
MAIN: nausea and vomiting
OTHER: colds, inflammation, pain
PARTS USED:
rhizomes

[6]-Gingerol

ABOVE **Gingerols are pungent compounds found in the fresh rhizomes of ginger (*Zingiber officinale*). On drying, they may be converted to the even more pungent shogaols.**

ABOVE **Ginger (*Zingiber officinale*) rhizomes are the part of the plant that will be most familiar, as they are used in cooking as well as medicinally.**

Ginger is no longer found in the wild but probably originally grew in tropical lowland forests in India and eastwards to south-central China. The medicinal use of ginger rhizomes (underground stems) in Asia can be traced back to Ayurvedic texts dating to 2000 BCE, which reported that it aided digestion and provided relief for rheumatism and inflammation. In China, ginger is recorded in the *Shen Nong Ben Cao Jing* (*The Drug Treatise of the Divine Countryman*), the earliest extant Chinese pharmacopoeia (*c.* 25–200 CE) (see page 16). Both Ayurvedic and traditional Chinese medicine consider fresh and dry ginger rhizome to be separate drugs used for various conditions. In the Middle East and Europe, ginger was employed medicinally as early as Greek and Roman times, and by the Middle Ages it was used for conditions such as colds, chest infections, joint pain and digestive complaints. Today, it is used medicinally worldwide and is also a popular flavouring in food, drinks and sweets (see box).

Ginger rhizomes contain numerous pungent compounds. Of these, the gingerols are more abundant in fresh rhizomes, but on drying some of these degrade to shogaols, which are twice as pungent. These compounds have a warming effect when eaten due to their stimulation of heat-producing (thermogenic) receptors. They also have a carminative effect that reduces nausea and vomiting by helping to break up and expel intestinal gas, and assist stomach emptying. In recent years, studies on its use as a herbal

Ginger rhizomes are aromatic and have a pungent, lemony flavour. Fresh rhizomes are a common flavouring in Asian cooking, while the dried, powdered form is more usual in Europe and elsewhere. Biscuits, cakes and some drinks can be flavoured with ginger, and as one of the warming spices it is particularly associated with winter in colder climates. One of the earliest records of gingerbread being made in the shape of a person was for a banquet held by Queen Elizabeth I (1533–1603), sovereign of England and Ireland, although they were probably used as love tokens before then. The gingerbread man is still popular today.

preparation to reduce nausea and vomiting (antiemetic) suggest that ginger may provide relief from motion sickness (see page 86) and also ease nausea caused by chemotherapy.

CANNABIS DURING CHEMOTHERAPY

Nausea and vomiting are experienced by up to 70–80 per cent of patients undergoing chemotherapy. In the 1970s, patients reported that they experienced less nausea and vomiting if they had smoked marijuana from the cannabis plant beforehand. The psychoactive THC was found to have the antiemetic effect. Various synthetic derivatives of THC were developed and tested, including nabilone, which was subsequently authorized in the United States and Europe as a pharmaceutical drug. As with marijuana, nabilone has been found to be effective if given before chemotherapy starts. It may work by direct inhibition of the vomiting control mechanism in the central nervous system (CNS).

TOP LEFT Illustration of ginger (*Zingiber officinale*), showing a rhizome from which arise upright pseudostems. The latter are either long and leafy, or shorter and bear a head of flowers.

LEFT In many parts of the world it is illegal to grow cannabis (*Cannabis sativa*) unless you have a special licence. Plants occasionally appear in gardens from dropped bird seed.

IN SICKNESS AND IN HEALTH

THE MIDDLE EAST

The Middle East is centred on the Arabian Peninsula, with Egypt in the west, Turkey in the north and Iran in the east. In ancient times it had an important role in trade thanks to its position at one end of the Mediterranean, its access to the Indian Ocean and overland routes to Africa and Asia, and the resources of the region itself. Often referred to as the cradle of civilization, the Middle East has had an influence far beyond its borders.

Biblical connections

In the book of Genesis of the Old Testament, Adam and Eve are thrown out of the Garden of Eden after being tempted to take a bite from a 'forbidden fruit'. Due to the association of Eden with the Middle East, historians have suggested that one of two plants native to the region is this mythical fruit. Fig (*Ficus carica*) in the mulberry family (Moraceae) is a candidate, partly because Adam and Eve used leaves from the tree to cover themselves after they had eaten the fruit; the other is pomegranate (*Punica granatum*) in the pomegranate family (Lythraceae). Both fruits have been used symbolically to represent fertility, but while the pomegranate has traditionally been used medicinally as a contraceptive, the fig has been most widely used as a mild laxative.

The Arabian Peninsula is also the source of two tree resins whose value is mentioned in the New Testament. Frankincense (*Boswellia sacra*) is native to Oman and Yemen, while myrrh (*Commiphora myrrha*) is also found in Saudi Arabia and East Africa; both trees are members of the frankincense and myrrh family (Burseraceae). The resins are traditional components of aromatic incense, in addition to having medicinal uses in both European and Asian systems of medicine.

Shifting influences

The use of medicinal plants in the early civilizations of Mesopotamia and Egypt were recorded in writing and depicted on the walls of tombs as early as 2700 BCE (see page 12). Subsequently, the Graeco-Roman medical

RIGHT *Qânûn fi'l tibb* (*Canon of Medicine*), volume 5, by Ibn Sina (Avicenna), Iran or Iraq, dated AH 444 or AD 1052, watercolour and ink on paper, Aga Khan Museum, Toronto, Canada.

traditions (see page 14) and trade with Africa and Asia influenced the region. Spices in particular were valued, both for their culinary and medicinal uses and as a symbol of prestige. Plant remains found at Quseir al-Qadim, an ancient port located on the Red Sea coast of Egypt, provide evidence for trade with Asia in spices and food plants during the period of Roman rule (1–250 CE). Black pepper (*Piper nigrum*) in the pepper family (Piperaceae) and beleric/belleric myrobalan (*Terminalia bellirica*) in the bushwillow family (Combretaceae) were among the species being imported during that period (see also page 45).

During the seventh century CE, the Arabs who lived in the Arabian Peninsula united under the leadership of Muhammad (c. 570–632 CE), the founder of Islam, and eventually conquered the entire region. Scholars translated Greek and Latin medical texts into Arabic, and in some instances these translations are the only surviving copies of the originals. In around 1020, the Arab physician Ibn Sina (Avicenna) (c. 980–1037) wrote the five-volume *Qânûn fi'l tibb* (*Canon of Medicine*), which was influenced by the Greek physician Galen of Pergamon (c. 129–200 CE; see page 14). It became the basis for the system of medicine known as Unani or Unani-Tibb, which was also influenced by the Indian Ayurvedic system of medicine. Unani is still practised in Muslim countries in central Asia and in parts of India (see page 44).

LEFT Frankincense, also known as olibanum, is the fragrant dried resinous sap of the frankincense tree (*Boswellia sacra*), which exudes from cuts made in the stems. It is highly valued in the perfume industry.

THE TREE OF LIFE

The tree of life is a traditional Middle Eastern motif. Images of the tree have often been identified as the date palm (*Phoenix dactylifera*) in the palm family (Arecaceae), which has a native range extending from the Arabian Peninsula east to Pakistan. The sweet fruit are highly valued in the region and the stems can also be tapped for their sap, which is drunk fresh or fermented. The tree of life is often depicted with side branches or a more complex arrangement, and it has been suggested that these are, in fact, a series of buds and blossoms of the Egyptian lotus (*Nymphaea nouchalii* var. *caerulea*, syn. *N. caerulea*) in the waterlily family (Nymphaeaceae). The Egyptian lotus was a symbol of the sun's cycle, as its blue flowers open in the morning to reveal the golden centre, like the sun in the sky, and close in the evening.

ABOVE Egyptian lotus or blue lotus (*Nymphaea nouchalii* var. *caerulea*) carries its flowers above the water surface.

Sick of moving

Many of us will have experienced motion sickness when travelling by car, plane or boat, or on a fairground ride. In all these cases our body is accelerating in unfamiliar ways, and the visual cues don't match up with the messages relating to balance. One effective treatment currently available is hyoscine (also known as scopolamine), an alkaloid found in some plants in the potato family.

MAGIC IN THE BELLY

PLANT:
Hyoscyamus niger L.

COMMON NAME(S):
black henbane

FAMILY:
potato (Solanaceae)

ACTIVE COMPOUND(S):

NATURALLY OCCURRING:
tropane alkaloids (hyoscine
[scopolamine], hyoscyamine)

SEMI-SYNTHESIZED:
hyoscine hydrobromide

MEDICINAL USES:

MAIN: motion sickness, to dry up secretions

OTHER: topical pain relief, fever, respiratory illness, nervous disorders

PARTS USED:
seeds, leaves, whole plant

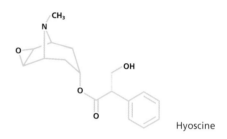

ABOVE **Hyoscine (scopolamine) is a tropane alkaloid found in certain members of the potato family (Solanaceae).**

BELOW **Black henbane (*Hyoscyamus niger*) is a biennial herbaceous plant with sticky, toothed leaves and cream flowers that usually have purple to brown venation. The appearance of its flowers and foul smell resemble rotting flesh and attract flies, which are its main pollinators.**

Black henbane is the most widely distributed species in the genus *Hyoscyamus*, growing in temperate Eurasia and northwest Africa, and also the most commonly used in traditional medicine. The oldest medical text to mention henbane, the Ebers Papyrus, an Egyptian manuscript dating to 1500 BC, recommended it for 'magic in the belly' and was probably referring to Egyptian henbane (*Hyoscyamus muticus*).

Henbanes produce tropane alkaloids that in humans have anticholinergic (antimuscarinic) effects on the CNS and peripheral nervous system, with the types and proportions of alkaloids in the plant determining which effects are more pronounced. In studies, black henbane was found to have more sedative effects than belladonna, also known as deadly nightshade (*Atropa bella-donna*; see page 180), which also contains tropane alkaloids, and so it became the focus of research. In 1880, the German pharmacist Albert Ladenburg (1842–1911) isolated a compound he called hyoscine, which differed from the previously isolated tropane alkaloid atropine only by the addition of a single oxygen atom. At around the same time, the closely related Japanese henbane (*Scopolia japonica*) and Russian henbane (*S. carniolica*) were also found to contain tropane alkaloids, one of which was named scopolamine. In the early 1890s, hyoscine and scopolamine were discovered to be the same compound.

SUPRESSING SIGNALS

Hyoscine reduces the mismatch in visual and balance signals received by the brain – the cause of motion sickness – by suppressing nerve signals from the middle ear that are associated with balance. In addition, it facilitates the habituation processes by which the body becomes used to the movement. Hyoscine is taken (as a hydrobromide derivative) either in tablet form, or is applied as a transdermal patch, which exploits its ready absorption through the skin. Historic reference to this quality is seen in the inclusion of henbane and deadly nightshade in the ointments that witches are said to have applied to their skin and that led them to experience the sensation of flying.

There are a number of other medical applications for hyoscine based on its anticholinergic action of drying up secretions, including as a premedication for inhalant halothane anaesthetics. In palliative care it is given to reduce excessive respiratory secretions, and its spasmolytic action can relieve bowel colic and the pain associated with it (see also page 90).

LEFT Egyptian henbane (*Hyoscyamus muticus*) is native to a region extending from northeast Africa east to India. It has a higher tropane alkaloid content than black henbane (*H. niger*) and has been collected from the wild as a commercial source of these compounds.

Devil's breath

In addition to sedation, hyoscine can induce a lack of will and amnesia. From at least medieval times, this effect was exploited during childbirth, giving rise to the term 'twilight sleep'. For a brief period in the early twentieth century, the practice was popular in the United States. The administration of injections of hyoscine and morphine enabled women to wake up after the birth with no memory of the pain of labour. These women also became more suggestible, leading to the use of hyoscine as a 'truth drug'. In larger doses, hyoscine can be lethal, a fact that has been exploited by poisoners throughout history. Hawley Harvey Crippen (1862–1910), known as Dr Crippen, was one of the most notorious of these, using the drug to poison his wife.

ABOVE Hyoscine from angel's trumpets (*Brugmansia* spp.) is known as devil's breath in South America, where it is used by robbers to render their victims unconscious – many thousands of such incidents are reported each year in Colombia alone.

Stomach ulcers

Damage to the lining of the stomach and duodenum can result in painful ulcers. There are several causes, including infection by the bacterium *Helicobacter pylori* and as a side effect of taking painkillers classed as non-steroidal anti-inflammatory drugs (NSAIDs). While plant compounds are not the basis for the modern pharmaceutical treatments for ulcers, herbal preparations may ease the symptoms and perhaps aid healing.

Glycyrrhizin

ABOVE **Glycyrrhizin is said to be at least 50 times sweeter than the sugar sucrose and can mask bitter flavours.**

ABOVE **Illustration of common liquorice (*Glycyrrhiza glabra*) including its woody rhizome, the part of the plant that is used medicinally.**

SWEET ROOT

PLANT:
Glycyrrhiza glabra L.

COMMON NAME(S):
common liquorice

FAMILY:
legume (Fabaceae)

ACTIVE COMPOUND(S):

NATURALLY OCCURRING:
triterpene saponins (including glycyrrhizic acid, also known as glycyrrhizin)

SEMI-SYNTHESIZED:
carbenoxolone sodium

MEDICINAL USES:

MAIN: gastric ulcers

OTHER: infections of the upper respiratory tract, inflammation

PARTS USED:
roots, rhizomes

At least six species of liquorice are used medicinally, with common liquorice (*Glycyrrhiza glabra*) probably the mostly widely adopted. Two species from China, Manchurian liquorice (*G. uralensis*) and Chinese liquorice (*G. inflata*), are authorized sources of herbal medicines in Europe, the United States and some Asian countries. The scientific name *Glycyrrhiza* is a combination of the Greek words *glykos*, meaning 'sweet', and *rhiza*, meaning 'root', and it is the sweet-tasting 'roots' – consisting of rhizomes (underground stems) and true roots – of these herbaceous plants that are employed both as a medicine and in confectionary (see box).

The principal sweet compound in liquorice root is the triterpene saponin glycyrrhizin. This is broken down in the intestines to yield glycyrrhetic acid, before being absorbed into the bloodstream. Glycyrrhetic acid is mildly anti-inflammatory and has been suggested as an aid in the healing of gastric ulcers. As early as the late 1940s, when the use of liquorice to treat ulcers was being investigated in the Netherlands, side effects were observed, including water retention, increased excretion of potassium and high blood pressure. In fact, although liquorice is generally regarded as safe, similar effects have been observed in people who consume large amounts in food or drinks.

Full of flavour

Liquorice isn't to everyone's taste, but it can be found in confectionary, drinks and tobacco products. Liquorice-based drinks can be traced all the way back to ancient Egypt, and their thirst-quenching properties aided soldiers and others undertaking long marches; they are still valued today in hot countries.

Common liquorice is native to southern Europe and the Middle East, eastwards to Mongolia and southwestern Russia. Its roots and rhizomes are used in the medicinal systems of all these regions and also widely elsewhere, including the Americas, it being one of the plants that accompanied early European colonizers. Traditionally, liquorice was taken to soothe the stomach, for coughs and congestion, to relieve mouth ulcers and as a tonic. In traditional Chinese medicine, liquorice root is a medicinal plant in its own right but is also used in the processing of other plants to minimize their potential side effects. Its sweetness can also help to mask the flavour of bitter herbal drugs.

ABOVE Water is used to make an extract from the roots and rhizomes of common liquorice (*Glycyrrhiza glabra*) and a few other related species. Following evaporation, a dark brown solid forms that is then used in confectionery and other products.

A REMOVED REMEDY

Glycyrrhizin and its derivatives have been identified as the cause of both the beneficial and adverse health effects of liquorice roots. A semi-synthetic derivative, carbenoxolone sodium, was used as an ulcer treatment in Europe in the 1960s but has now been largely replaced by other drugs due to its adverse effects. To reduce the risk of side effects, glycyrrhizin is now removed from some liquorice root herbal preparations. There is, however, little evidence that such preparations are effective at treating gastric ulcers. Glycyrrhizin is also used, or is under investigation for, other conditions – in China and Japan, for example, it is frequently given to protect the liver (hepatoprotective).

RIGHT The herbaceous Manchurian liquorice (*Glycyrrhiza uralensis*) grows in northern China and Tibet. It is used more frequently than common liquorice (*G. glabra*) in Asian systems of medicine.

When the bowel is irritable

Irritable bowel syndrome usually starts in early adulthood and can be a lifelong condition that affects up to 20 per cent of the population. It is characterized by pain or discomfort in the abdomen, bloating, flatulence and irregular bowel habits that fluctuate between constipation and diarrhoea. Today, plant-based pharmaceuticals derived from peppermint and corkwood (*Duboisia* spp.) are among the treatment options.

AFTER-DINNER MINTS

PLANT:
Mentha × piperita L.
COMMON NAME(S):
peppermint
FAMILY:
mint (Lamiaceae)
ACTIVE COMPOUND(S):
NATURALLY OCCURRING:
essential oil (includes the monoterpenes menthol and menthone)
MEDICINAL USES:
MAIN: irritable bowel syndrome
OTHER: digestive aid, headache, colds
PARTS USED:
oil, leaves, whole plant

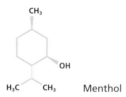

RIGHT **Menthol is the main compound in the essential oil from peppermint (*Mentha × piperita*).**

Menthol

ABOVE AND ABOVE RIGHT **Peppermint (*Mentha × piperita*) has the square stem of members of the mint family (Lamiaceae), toothed leaves in opposite pairs and heads of small mauve flowers.**

Peppermint is traditionally a symbol of purity and hospitality, and its fresh scent and taste will be familiar from sweets, chewing gum, ice cream and toothpaste. Growing to about 60 cm (24 in) tall, this herbaceous member of the mint family is a natural hybrid between spearmint (*Mentha spicata*) and water mint (*M. aquatica*), and was named in the eighteenth century from a specimen discovered in England. As a hybrid, it is itself sterile, relying on spreading runners to increase its population. Peppermint is considered to be native to Europe and central Asia, but is widely introduced elsewhere.

The leaves of species of mint (*Mentha* spp.) produce an essential oil that often contains menthol, with peppermint containing one of the highest menthol concentrations. Oil consisting of approximately 50 per cent menthol is extracted from fresh peppermint leaves by steam distillation. The cooling effect of menthol is through stimulation of 'cold sensation' receptors on neurons. It also relaxes the smooth muscle of the gastrointestinal tract by blocking calcium channels. If menthol is taken as an enteric-coated formulation, these actions are delayed until it reaches the intestines and bowel. Such preparations have been found to reduce the number and strength of gut contractions, and it is suggested that they also temporarily cause pain-sensing neurons to become less sensitive.

CORKWOOD EASES SPASMS

Hyoscine, a tropane alkaloid (see page 33), has some benefit in the treatment of IBS due to its action as a spasmolytic, calming gastrointestinal spasms. Such treatment is usually in the form of the butylbromide derivative of hyoscine. Due to its modified chemical structure, hyoscine butylbromide doesn't cross the blood–brain barrier and so is less likely to cause side effects in the CNS. Hyoscine is still obtained from plants in the potato family, with most of the supply coming not from henbanes (*Hyoscyamus* and *Scopolia* spp.; see page 86), but from an Australian tree. The corkwoods are a genus of only four species, two of which – *Duboisia myoporoides* and *D. leichhardtii* – are important for the hyoscine content of their leaves, with some hybrids between the two containing the highest concentrations.

Duboisia myoporoides was added to small water sources by Aboriginal Australians to poison fish and emu (*Dromaius novaehollandiae*), and was also used medicinally to inhibit bodily secretions and relieve allergies and colds. Hyoscine was first isolated from corkwoods in the late nineteenth century, but wider interest in it came about only during the Second World War. At that time there was an increased demand for hyoscine, principally for its use to prevent sea sickness in the Allied forces. Plant material was initially collected from the wild, but there are now plantations both in Australia and elsewhere.

ABOVE ***Duboisia myoporoides*** is a small tree from eastern Australia that can grow to 14 m (46 ft) tall. It can withstand fire due to its corky bark and will regrow from suckers if it is chopped down.

RIGHT **Peppermint (*Mentha* × *piperita*)** has been used traditionally as a digestive aid.

Soothing inside and out

Mint is recorded in Egyptian graves dating from 3200–2600 BCE, was used by the Romans, and is mentioned in medieval European texts. Today, fresh or dried peppermint leaves are taken as a tea or used as a flavouring in food, where the menthol acts on the stomach and has a mild, soothing effect. Trapped air can be released from the stomach due to relaxation of the muscles of the oesophageal sphincter, but this can exacerbate symptoms in those with gastro-oesophageal reflux disease. When peppermint was added to the London Pharmacopoeia in 1721, it was prescribed internally for conditions such as indigestion, flatulence and colic, and used externally for nervous headaches, fever and colds. These traditional uses of peppermint have stood the test of time.

Let's get things moving

Insufficient fibre or fluids in the diet, taking certain medicines such as opioid painkillers (see pages 97 and 140) or a lack of exercise can all lead to constipation. A number of plants come to the rescue, easing the situation through one of two means. Bulk-producing laxatives, taken with liquids, soften and expand the stools, stimulating bowel emptying. In contrast, plant laxatives containing anthranoid compounds, such as senna (*Senna alexandrina*), have a stimulant, or purgative, action.

SHORT-TERM SOLUTION

PLANT:
Senna alexandrina Mill.
(syn. *Cassia acutifolia* Delile,
Cassia angustifolia Vahl,
Cassia senna L.)
COMMON NAME(S):
senna, Alexandrian senna,
Tinnevelly senna, Indian
senna
FAMILY:
legume (Fabaceae)

ACTIVE COMPOUND(S):
NATURALLY OCCURRING:
dianthrone glycosides
(sennosides)
**SYNTHESIZED FROM NATURAL
LEAD:** dantron (danthron)
MEDICINAL USES:
laxative
PARTS USED:
leaves, fruit

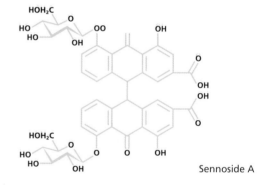

Sennoside A

ABOVE **Sennoside A is one of the main dianthrone glycosides in senna (*Senna alexandrina*), and contributes to the laxative action of both leaves and pods.**

ABOVE **Senna (*Senna alexandrina*) is found throughout northern Africa, from Mauritania in the west to Somalia, and through the Arabian Peninsula to Pakistan and India in the east. Long-term use can weaken the colon's spontaneous contractions, causing dependence on laxatives to avoid constipation.**

In the ninth century CE, camel caravans carried the leaves of wild senna plants (*Senna alexandrina*) from southern Egypt to the Nile or Red Sea ports, and Isaac Judaeus, a contemporary Egyptian writer, described their action as a laxative. Today, senna is still used for that purpose globally, as both herbal preparations and pharmaceutical drugs for short-term use.

While the genus *Senna* contains approximately 300 species, the common name 'senna' refers to just one of these species (see box opposite). Its leaves are harvested in the summer, and both the leaves and pods are harvested in the autumn. In fact, the growing conditions, the time of harvest and the plant part all affect the amount of anthranoid compounds in the plant material, so medicinal preparations are usually standardized.

BELOW The fruit of senna (*Senna alexandrina*), known as senna pods, have a milder laxative effect than the leaves.

The sennosides in the dried leaves and pods pass largely unchanged through the stomach and small intestines, but when they reach the large intestines they are broken down by bacteria into the active compounds, principally rhein anthrone. These stimulate the contractions of the smooth muscle of the colon (peristalsis), and inhibit the uptake of water by the cells lining the large intestine, resulting in softer stools and bowel emptying. A minor constituent of senna, dantron (danthron; 1,8–dihydroxyanthraquinone), is available in some countries as a synthesized drug. However, due to concerns that dantron poses a potential carcinogenic risk, its use is restricted.

SACRED BARK

Cascara sagrada (literally 'sacred bark') was the name given to *Frangula purshianus* (syn. *Rhamnus purshiana*), a North American shrub in the buckthorn family (Rhamnaceae), by Spanish explorers in the seventeenth century. They adopted the tradition of the Native Americans living in the species' natural west coast range of using bark aged for one year as a laxative, and their name shows its importance as a medicine. Dried fruits of buckthorn (*Rhamnus cathartica*) from Europe and northern Africa were used from at least the ninth century CE for the same purpose, as well as for reducing water retention and alleviating rheumatism.

The compounds found in fresh buckthorn material can stimulate vomiting and cause colic, and are strongly purgative. Ageing the bark and drying the fruit results in the oxidization of these compounds to anthraquinone glycosides (principally cascarosides in cascara sagrada), which are less likely to produce the unwanted effects in the small intestines. The cascarosides are metabolized by gut bacteria in the large intestines, mainly to emodin anthrone, which acts in the same way as rhein anthrone from senna.

RIGHT Cascara sagrada (*Frangula purshiana*) is a shrub or small tree growing to 12 m (40 ft) tall that is native to western North America and Mexico.

Alexandrian and Tinnevelly senna

Senna, a woody-stemmed perennial, has had many scientific names (see fact file), reflecting earlier beliefs that it was several species, not the single species *Senna alexandrina* that botanists recognize today. It also has many vernacular names, reflecting the countries from which it was collected and the routes along which it was traded. The name senna itself is derived from the Arabic word for the plant, *sena*. Alexandrian senna was considered to be the highest quality. It was collected from the wild, mainly in Egypt and Sudan, and shipped via the Egyptian port of Alexandria. Tinnevelly or Indian senna came from plants cultivated in wetter conditions in India and Pakistan. Since the nineteenth century, India has been the main country of production, but commercial sources of senna are also found in Egypt, Sudan and Malaysia.

Playing fast and loose

Diarrhoea is one of the body's primary defence mechanisms against the ingestion of harmful substances, so it comes as no surprise that there are many causes of the condition. While eating poisonous plants frequently results in diarrhoea, many plants are traditionally used to ease the condition. Some of the most successful are seirogan (which contains beechwood creosote) and pharmaceutical opiates (see page 96), as well as dragon's blood.

Bernabé Cobo (1582–1657), a Spanish Jesuit missionary working in Peru and Bolivia during the seventeenth century, recorded the medicinal uses of *sangre de grado*, 'dragon's blood' (*Croton* spp.), including for 'all kinds of incontinence' and wounds. Cobo is probably better known for his part in introducing to Europe the bark of cinchona (*Cinchona* spp.; see page 134), the source of an important anti-malarial. Several centuries later, an extract from the latex of dragon's blood is now a drug approved by the United States Food and Drug Administration (FDA) for the treatment of diarrhoea in people with HIV/AIDS, and it is used elsewhere for 'traveller's' diarrhoea and diarrhoea associated with cholera.

Croton is one of the largest plant genera, with around 1,185 species found throughout the tropical and subtropical regions of the world. The common name dragon's blood is used for a few *Croton* species from western South America that produce a distinctive red latex. In the 1990s, investigations revealed the active constituents of the latex to be a mixture of proanthocyanidins, known collectively as crofelemer, with different numbers of catechin sub-units linked into polymer chains. Crofelemer is poorly absorbed by the gastrointestinal tract, but acts on it directly by targeting the chloride channels in gut cells that are involved in the movement of water into the bowel.

SHARING BENEFITS?

Croton lechleri was considered to be the most suitable source for crofelemer as it is a fast-growing pioneer species that can tolerate a range of conditions. The latex is obtained by cutting down and draining whole

RIGHT **Dragon's blood (*Croton lechleri*) has broad leaves and reaches 10–15 m (33–50 ft) in height. The tree grows in lowland forest of the northwest Amazon, particularly on disturbed ground and along rivers and streams.**

DRAGON'S BLOOD

PLANT:
Croton lechleri Müll.Arg.
COMMON NAME(S):
dragon's blood, sangre de grado, sangre de drago
FAMILY:
spurge (Euphorbiaceae)
ACTIVE COMPOUND(S):
NATURALLY OCCURRING:
proanthocyanidins (crofelemer)

MEDICINAL USES:
MAIN: anti-diarrhoeal
OTHER: externally for wounds
PARTS USED:
latex

trees once they are six or seven years old. Local knowledge is used to select trees, and the best time of day and weather, for the greatest harvests. The crop can be sustainable if several trees are planted for each one that is removed. Some companies working in Peru also provide fair pay to employees for collecting the latex and replanting trees, with additional support to communities for health and education

As 2012 turned into 2013, crofelemer became only the second drug to get FDA approval through the 'botanicals' route, the first being an extract of green tea (*Camellia sinensis*; see page 25). In 2018, the manufacturers of crofelemer signed an agreement with the AIDS Drug Assistance Programs to make the drug available cheaply to people in the United States on low incomes. Some argue that the original holders of the traditional knowledge have themselves seen little return from the hundreds of millions of pounds poured into investment and sales. Due to the scale of use locally and interest worldwide in dragon's blood for pharmaceutical and other products, it is one of the plants prioritized by the Peruvian government for protection against biopiracy.

Sealing and healing

The striking red latex of dragon's blood is traditionally used in Amazonia for conditions of the gastrointestinal tract and circulation, including ulcers of the mouth and stomach, tonsillitis, diarrhoea and blood in the urine. Externally, it was used to seal open wounds and prevent their infection, and as a wash to stop bleeding from haemorrhoids. Preparations for wound healing, bites and stings are available elsewhere, including the United States, and are also used in cosmetics.

ABOVE **Dragon's blood latex (also referred to as a resin) in its dried form and as a liquid extract. In addition to its medicinal uses, the latex is now of interest for use in cosmetics.**

LEFT **It was originally hoped that the latex in dragon's blood (*Croton lechleri*) could be tapped in the manner of tapping latex from the rubber tree (*Hevea brasiliensis*), also in the spurge family. The ducts that carry the latex (lacticifers) are different in the two species, however, with those of dragon's blood unable to regenerate once cut, unlike rubber lacticifers.**

EFFICACIOUS DEWDROPS

PLANT:

Fagus sylvatica L. and
Fagus crenata Blume

COMMON NAME(S):

European beech; Japanese
beech

FAMILY:

beech (Fagaceae)

ACTIVE COMPOUND(S):

NATURALLY OCCURRING:

phenols (guaiacol, creosol,
phenol, cresol)

MEDICINAL USES:

MAIN: dysentery, diarrhoea

OTHER: expectorant,
antiseptic

PARTS USED:

wood resin

BELOW **The Japanese beech (*Fagus crenata*) is found widely in the
cool temperate forests of Japan, where it can be the dominant tree
species and grows to 35 m (115 ft) in height.**

Guaiacol

ABOVE **Guaiacol is a colourless to yellow liquid with a characteristic
smoky odour, and is one of the primary compounds behind the taste
of whisky. In Japan, medicinal preparations using guaiacol must be
standardized to contain between 23 and 35 per cent of the phenol.**

It has been a stark reality throughout history that,
during times of war, more soldiers die from disease
and hardship than are killed in battle. It is no
surprise then that when they embarked on a campaign
against Russia in 1904, the Japanese looked for ways to
prevent one of the biggest causes of debilitation and
death: dysentery. There are different stories about how
they came to use wood creosote, but whatever the
truth, the commercial success of seirogan (as it came
to be known) is unquestionable. Originally, the name
seirogan was represented by characters that translated
as 'conquer Russia pill', a feat of patriotic marketing.
While pronunciation of the name has remained the
same over time, the characters have since been
replaced by those for 'efficacious dewdrops' and the
original meaning is all but forgotten.

The Japanese Pharmacopoeia, which lists medicinal
substances used in Japan, includes an entry for wood
creosote, or *creosotum ligni*, a mixture of phenols such
as guaiacol that are obtained by dry distillation from
wood tar followed by further purification processes.
Branches and stems from different tree genera can be
the source of the wood tar, including species of beech
(*Fagus* spp.) in the beech family (Fagaceae). The anti-
diarrhoeal action of wood creosote phenols involves a
reduction of water secretion into the intestines, and
a reduced rate of muscle contraction in the colon. In
contrast to crofelemer (see page 94), phenols such as
guaiacol are absorbed from the stomach and circulated
in the blood, and have an effect on the gastrointestinal
system in as little as 15 minutes.

MAKING THE MOST OF A SIDE EFFECT

The opium poppy is probably best known for being a
source of the opiate painkillers morphine and codeine
(see page 140). One of the usually less desirable effects
of opiates is constipation. The drugs activate opioid
receptors in the nervous system of the digestive
tract, the enteric nervous system. This results in
physiological effects, including inhibition of stomach
emptying, the movement of food through the

intestines and the secretion of fluid into the intestines. In combination, these lead to constipation, and although this is unwelcome if the aim of taking opiates is pain relief, it can be used positively to treat diarrhoea. Morphine has been used in anti-diarrhoea preparations such as in 'kaolin and morphine'. More recent developments are synthetic opioid drugs such as loperamide, which are only poorly absorbed by the gastrointestinal tract and so their effects are largely limited to the enteric nervous system, making them effective as a treatment for diarrhoea.

LEFT Beechwood creosote can be derived from several tree genera other than beech (*Fagus*), namely pine (*Pinus*) in the pine family (Pinaceae), Japanese sugi pine (*Cryptomeria japonica*; shown here) in the swamp cypress family (Taxodiaceae), *Afzelia* (syn. *Intsia*) in the legume family, *Shorea* in the dipterocarp family (Dipterocarpaceae) or teak (*Tectona*) in the mint family.

Sources of creosote

There are different types of creosote and it is important not to confuse them as they vary chemically and in terms of their toxicity. For clarity, wood creosote used as an anti-diarrhoeal is often referred to as beechwood creosote. It was first extracted in 1832 by Baron Dr Carl Ludwig von Reichenbach (1788–1869) through dry distillation of wood from the European beech (*Fagus sylvatica*). Reichenbach found that beechwood creosote could be used to preserve meat, and its antiseptic properties were later used medicinally, including for tuberculosis.

The term wood creosote is also used for the resin derived from the leaves of the creosote bush (*Larrea tridentata*) in the caltrop family (Zygophyllaceae). Leaves from this species are the source of a herbal preparation called chaparral, whose prolonged use has been reported to cause liver damage. There is therefore concern that similar effects might result from the use of creosote bush resin, which contains nordihydroguaiaretic acid (NDGA).

The third type of creosote is coal tar creosote, which is not derived from living plants but from their long-decayed remains. Coal tar creosote is

an oily liquid consisting mainly of polycyclic aromatic hydrocarbons and is widely used as a wood preservative. Along with its source, coal tar, coal tar creosote has been found to be carcinogenic, with effects resulting from extended exposure via the skin and respiratory system.

ABOVE European beech (*Fagus sylvatica*) is found throughout Europe as far east as the Caucasus. It is a large tree that can reach more than 40 m (130 ft) in height. Its nuts, called beech mast, are a food source for badgers (*Meles meles*), squirrels and birds, especially finches.

IN SICKNESS AND IN HEALTH

Intestinal worms

Intestinal worms and other parasites of the digestive system can cause diarrhoea, loss of appetite, anaemia, reduced body weight and, eventually, death if left untreated. Numerous plants have been employed to try to control such infestations, but there can be a fine line between a dose sufficient to kill or expel the worms and an amount that would seriously harm the patient. Safer synthetic compounds are now therefore adopted instead.

IT'S A DOG'S LIFE

PLANT:

Areca catechu L.

COMMON NAME(S):

betel palm, areca palm

FAMILY:

palm (Arecaceae)

ACTIVE COMPOUND(S):

NATURALLY OCCURRING:

reduced pyridine alkaloid
(arecoline)

SYNTHESIZED FROM NATURAL

LEAD: arecoline
hydrobromide

MEDICINAL USES:

MAIN: veterinary anthelmintic

OTHER: psychoactive

PARTS USED:

seeds

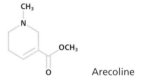

Arecoline

ABOVE **Betel nuts, the seeds of the betel palm (*Areca catechu*), contain the alkaloid arecoline.**

ABOVE **Coloured scanning electron microscope image of arecoline crystals.**

The slender betel palm (*Areca catechu*) is thought to originate from the Philippines, but is now found quite widely in wet tropical and coastal regions of the world thanks to the use of its nuts (seed kernels) as a stimulant, to alleviate fatigue and suppress the appetite. An estimated 600 million people worldwide use the seeds of the betel palm for their euphoric effects. The active substance, the alkaloid arecoline, is absorbed from the mouth, as well as from the intestines if swallowed. Like pilocarpine from jaborandi (*Pilocarpus jaborandi*; see page 178) in the citrus family (Rutaceae), betel nut stimulates cholinergic (muscarinic type) receptors in the nervous system. In addition to mild euphoria, symptoms include dizziness, salivation, pupil contraction and tremor.

Betel nuts and, later, the hydrobromide derivative of arecoline were used in the nineteenth and twentieth centuries to deworm animals, including dogs with tapeworm. This treatment exploited one of arecoline's effects, which is to increase muscle contractions in the intestines, giving it a purgative action. Purging (severe diarrhoea) usually takes place within two hours, expelling the worms, with vomiting a frequent side effect. While the use of arecoline hydrobromide to treat tapeworms in dogs had declined by the 1960s, it continued to be a useful tool to assess the prevalence of the parasites in dog populations in many countries around the world. That application has also now been abandoned in favour of drugs that are more effective and better tolerated.

ABOVE **Betel palm (*Areca catechu*) bearing ripe fruit. Each fruit contains a single seed.**

WORMWOOD AND MALE FERN

The decline in the use of plant-derived compounds for the treatment of worms in humans can be illustrated by some examples that fell out of favour in the United States in the twentieth century. Santonin, from the flower heads (capitula) of Levant wormseed or wormwood (*Seriphidium cinum*, syn. *Artemisia cina*) from Russia, Turkestan and the southern Urals, and from sea wormwood (*Artemisia maritima*), both in the daisy family (Asteraceae), was used to treat roundworms. Aspidium oleoresin, an extract from the rhizome of the male fern (*Dryopteris filix-mas*, syn. *Aspidium filix-mas*) consisting of phloroglucinol derivatives, was the active ingredient in an eighteenth-century remedy called Madame Nauffer's Tapeworm Cure, and was also used for liver flukes. Both santonin and aspidium oleoresin were removed from the *United States Dispensatory* in the second half of the twentieth century. Another phloroglucinol derivative, desaspidin, from the rhizomes of a few other species of *Dryopteris*, including mountain wood fern (*Dryopteris campyloptera*, syn. *D. austriaca*), was found to be a better taenicide (tapeworm killer) than aspidium oleoresin. With parasites now becoming resistant to synthetic treatments, humans are once more looking to natural sources for novel compounds, and perhaps these plants will again prove useful.

Chewing betel nuts is usually a social habit and in any case is difficult to do secretly, as both saliva and teeth become stained red. It is used in the form of a quid or wad made from betel nuts and leaves of the betel vine (*Piper betle*) in the pepper family (Piperaceae), together with calcium hydroxide (also known as slaked lime). Other substances, such as tobacco and spices, can also be added. Traditional reasons for betel chewing are to aid digestion, as a prophylactic against dysentery, and to endure hunger and thirst. However, the practice is not without its risks; even without the presence of tobacco it is addictive and has been associated with an increased risk of cancer of the larynx and oesophagus.

ABOVE **Young dancer in Papua New Guinea with teeth stained red from chewing betel nut. Use of betel nut can be recreational, social, or associated with celebrations and ceremony.**

CHAPTER 5

SUPPORTING THE ORGANS AND GLANDS

Organs, glands and even cells within the body can become damaged or diseased, or cease to function normally. The liver, for example, is vital to the detoxification of harmful compounds but can itself be harmed in the process, and through normal age-related changes the prostate can hamper urination in men. This chapter focuses on herbal preparations that target liver disease and benign prostate conditions, and also looks at research into diabetes.

Milk thistle (*Silybum marianum*)

NATURAL COMPOUNDS FOR ORGANS AND GLANDS

Functions within the human body are undertaken by specialized cells organized into tissues, organs and systems. This chapter looks at diseases and conditions that affect some of these and the plants that have been explored to treat them. Many of the plants described have evidence only from traditional use, which may be supported by research into mechanisms of action. Others have resulted in the development of pharmaceutical drugs based on single compounds.

LIVER TONICS

The vital functions performed by the liver, one of the largest of our organs, mean that it comes into contact with harmful compounds and pathogens that can damage it, leading to a number of diseases. Plants have been used traditionally to protect the liver from damage, including milk thistle (*Silybum marianum*; see page 104) in the daisy family (Asteraceae). A pharmaceutical developed from one of its active compounds, silibinin, is approved in some countries for the treatment of poisoning from mushrooms that contain liver-damaging (hepatotoxic) amatoxins. For other plants, recent studies have looked into mechanisms that might infer an ability to protect or treat the liver, but clinical data are lacking. Fruit of the Chinese magnolia vine (*Schisandra chinensis*; see page 106) in the magnolia vine family (Schisandraceae) are eaten as a food and also used in traditional Chinese medicine as a general tonic, and for liver disease. The rhizomes of turmeric (*Curcuma longa*, syn. *C. domestica*; see page 106) in the ginger family (Zingiberaceae), originally from India, are widely used both as a spice and medicinally. Traditionally, turmeric is also considered to be a digestive aid.

THE AGEING PROSTATE

The prostate is a gland in the male reproductive system, positioned around the urethra at the neck of the bladder. As men age, it increases in size and can restrict the diameter of the urethra, resulting in the

LEFT Turmeric (*Curcuma longa*) plants are cultivated in Asia, Central and South America, and elsewhere. Rhizomes from different forms of turmeric are used fresh or dried and powdered as a spice, for colouring and medicinally.

need to urinate more frequently, reduced flow rate and inability to empty the bladder completely. When there is no underlying disease associated with prostate enlargement, the condition is called benign prostatic hyperplasia (BPH). Herbal preparations and dietary supplements for BPH that are based on traditional use include the saw palmetto (*Serenoa repens*; see page 110) in the palm family (Arecaceae).

TOO MUCH SUGAR

The pancreas is an organ in the abdomen that has two functions. As part of the digestive system it produces enzymes that help to break down food as it enters the small intestines. In addition, it manufactures the hormone insulin, which regulates the uptake of the sugar glucose by cells in the body. Type 1 diabetes (also called insulin-dependent diabetes) occurs when the immune system attacks the cells in the pancreas and they cease to produce insulin. Without insulin, levels of glucose in the blood rise and cells are starved of glucose for energy. This should be treated by regular injections of insulin, although this isn't always possible where funds or treatment facilities are unavailable.

Type 2, or non-insulin-dependent, diabetes is much more common and is on the increase worldwide. It usually occurs in later life and can be associated with diets high in processed carbohydrates, among other

ABOVE **Asiatic or common dayflower (*Commelina communis*) is a prostrate annual that is native to East and Southeast Asia and far east Russia. Aerial parts are used in traditional Chinese medicine to 'clear heat', and in Korea to control blood sugar levels.**

risk factors. In this type of diabetes the pancreas is still producing insulin but the cells are not responding to it as efficiently (called insulin resistance), resulting in high blood glucose levels. Diet and exercise can help to overcome insulin resistance in some cases, but medication may also be needed. The most commonly used first-line treatment for type 2 diabetes, metformin, was developed from a compound found in goat's rue (*Galega officinalis*; see page 112) in the legume family (Fabaceae), although it was a long journey. The template for the anti-diabetic drug miglitol is an iminosugar, 1-deoxynojirimycin (1-DNJ), found in a number of bacteria and plants such as the Asiatic or common dayflower (*Commelina communis*; see page 114) in the spiderwort family (Commelinaceae). Many other plants are traditionally used as remedies for diabetes – including bitter gourd (*Momordica charantia*; see page 114) – even though clinical evidence to support this use is still lacking.

Loose liver

Cirrhosis (usually related to alcohol), fatty liver disease and viral hepatitis are the biggest categories of liver disease today. A variety of herbal products are used around the world with the aim of ameliorating their effects, including the Chinese magnolia vine and turmeric. First, we look at the milk thistle, which is used for a more unusual cause of liver damage.

MILK THISTLE

PLANT:
Silybum marianum (L.) Gaertn. (syn. *Carduus marianus* L.)

COMMON NAME(S):
milk thistle, holy thistle, Mary thistle

FAMILY:
daisy (Asteraceae)

ACTIVE COMPOUND(S):

NATURALLY OCCURRING:
lignans (silymarin – includes silibinin (silybin), silidianin

(silydianin), silicristin (silychristin))

SEMI-SYNTHESIZED:
silibinin-*C*-2′,3-dihydrogen succinate, disodium salt

MEDICINAL USES:
MAIN: amatoxin poisoning
OTHER: liver disorders

PARTS USED:
fruit

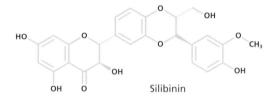

ABOVE **Milk thistle (*Silybum marianum*) fruit contain 1.5–3.0 per cent silymarin, which is a mixture of flavolignans containing approximately 50 per cent silibinin (silybin).**

ABOVE **Milk thistle (*Silybum marianum*) is an annual or biennial plant with tufts of tubular reddish-violet florets. The fruit (achenes), often referred to as 'seeds', are used medicinally to protect the liver.**

With its spiny, marbled white leaves, above which rise spiky heads of purple flowers, milk thistle is a striking plant. Although native to North Africa, the Mediterranean and central Asia, it is now widely naturalized elsewhere and is also cultivated around the world as a medicinal plant. The medicinal part is the fruit, which contains a mixture of lignans known as silymarin. Milk thistle fruit have been used for liver complaints for more than 2,000 years (see box), and they are now one of the most popular herbal preparations as we look for ways to counteract our lifestyle choices or to support conventional treatment for liver conditions. Clinical studies into the usefulness of taking herbal milk thistle have had mixed results, which could in part be due to differences in the doses taken. However, there is some experimental evidence for liver protection and regeneration effects.

DEATH CAP AND DESTROYING ANGEL

Foraging for edible mushrooms is a traditional activity in many countries and a pleasant pastime in others. If care isn't taken to ensure that the identity of every picked mushroom is known, one of the poisonous species can enter the mix, sometimes with serious consequences. The most poisonous of the compounds found in mushrooms are cyclopeptides, particularly amatoxins such as amanitin, which are damaging to the liver and other organs. They are found in several species of the *Amanita* genus, including death cap (*A. phalloides*) and the destroying angel (*A. virosa*), as well as in some members of *Lepiota* and *Galerina*. Together, these species are responsible for most mushroom fatalities worldwide. Aggressive treatment, including liver transplants, has reduced the number of deaths in more recent years, and the use of a preparation from milk thistle, particularly in Europe, has contributed to this success. Unlike the herbal preparations from milk thistle, which are extracts taken by mouth, a water-soluble derivative of silibinin is given intravenously to treat amatoxin poisoning. Silibinin's action includes inhibiting the transport of amatoxins across the membrane of liver cells (hepatocytes). Intravenous administration of a silibinin derivative is also showing promising results against hepatitis C.

LEFT **Poisoning from death cap (*Amanita phalloides*) starts with effects on the digestive system five to twenty-four hours after the mushroom has been eaten. These improve with symptomatic treatment, and it is not until two to four days later that the more severe symptoms of liver and kidney toxicity become evident.**

A holy connection

It is said that milk from the Virgin Mary splashed onto the leaves of milk thistle, which is why they are marbled with white. This association is recorded in many of the plant's common names, including holy thistle, and in its scientific species name, *marianum*. In fact, one of milk thistle's traditional uses was to stimulate milk production in new mothers, and it was also used to soothe coughs and colds, bloating and flatulence, menstrual complaints and painful joints. The Roman naturalist Gaius Plinius Secundus, known as Pliny the Elder (23–79 CE), recorded that the juice of milk thistle mixed with honey would 'carry off bile'. The traditional use of the plant for liver disorders is probably the most widespread, being found in European and Asian systems of medicine.

In addition to its medicinal uses, most parts of milk thistle are edible. Young leaves, stems and roots are cooked as a vegetable. The base of the unopened flower head is eaten like a globe artichoke (*Cynara cardunculus* var. *scolymus*; also in the daisy family and of interest for its potential liver-protectant effects), and the fruit (kenguel seeds) can be roasted, ground and used as a replacement for coffee.

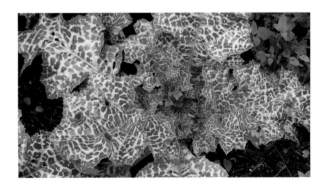

ABOVE **According to the doctrine of signatures, the marbled leaves of milk thistle (*Silybum marianum*) indicated its use to stimulate milk production in new mothers.**

CHINESE MAGNOLIA VINE

PLANT:

Schisandra chinensis (Turcz.)
Baill.

COMMON NAME(S):

Chinese magnolia vine,
schisandra

FAMILY:

magnolia vine
(Schisandraceae)

ACTIVE COMPOUND(S):

NATURALLY OCCURRING:

lignans (schisandrins,
gomisins); triterpenes
(nigranoic acid)

MEDICINAL USES:

MAIN: adaptogen, and to
protect the liver and improve
liver function

OTHER: tonic

PARTS USED:

fruit

The Chinese magnolia vine is a woody climber reaching up to 8 m (26 ft) high and is native to north China, the Korean Peninsula, far eastern Russia and central Japan. Its edible fruit (see box) are used medicinally in those countries and also in Europe and North America. Chinese magnolia vine fruit are described as an adaptogen, in the sense that they are a general tonic, able to normalize body functions and strengthen systems compromised by stress. They are therefore used as a herbal medicine, sometimes in combination with ginseng (*Panax ginseng*; see page 64), for a great variety of conditions affecting many body organs and systems. Notable among these are remedies for the liver, which claim to improve liver function, including in people with hepatitis.

A number of laboratory studies have looked into the active compounds of Chinese magnolia vine fruit and their potential mechanisms. It has been found, for example, that the lignans in the fruit are cytochrome P (CYP) transporter inhibitors. As one cause of liver injury is the activation of chemicals such as paracetamol (acetaminophen) by CYP, it is thought that the inhibition of CYP contributes to Chinese magnolia vine's hepatoprotective effects. However, some pharmaceutical drugs rely on activation by CYP to exert their beneficial effects, and others are deactivated by CYP, potentially reducing their effectiveness. There is the potential, therefore, for Chinese magnolia vine either to enhance or diminish the activity of other drugs.

STAINING YELLOW

Turmeric is another plant with a long history of traditional use for disorders of the liver, among other conditions. Its rhizomes resemble those of the closely related ginger (*Zingiber officinale*; see page 82) but its fresh flesh is a vivid orange or yellow colour – the Chinese Pinyin name for the rhizomes is *jiang huang*, literally 'ginger yellow'. They are used fresh or dried and powdered, and give foods – including many dishes from India and prepared mustard – their distinctive colour. The main pigment is curcumin, which is considered to be the principal medicinally active compound in the rhizomes. Originating in India, turmeric is used in the Ayurvedic, Siddha and Unani systems of medicine, as well as elsewhere in Asia and now also worldwide. It is thought to stimulate the liver to produce bile (choleretic), which aids in the breakdown of fat during digestion. Laboratory studies

LEFT **Chinese magnolia vine (*Schisandra chinensis*) grows in lowland ravine forests and along rivers. The fruit, which can be up to 7.5 mm (⅓ in) diameter, are carried in long bunches.**

suggest curcumin may protect the liver from the damage caused by harmful substances, including alcohol, although the amount in your curry is unlikely to be sufficient to counteract any overindulgence. Turmeric is also being investigated for use in inflammatory diseases and for its potential anticancer properties (see page 216).

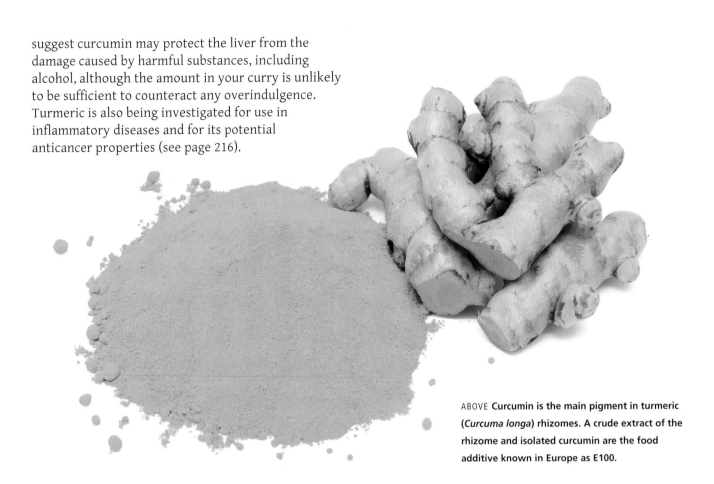

ABOVE Curcumin is the main pigment in turmeric (*Curcuma longa*) rhizomes. A crude extract of the rhizome and isolated curcumin are the food additive known in Europe as E100.

Five-flavoured fruit

The Chinese Pinyin name for the fruit of the Chinese magnolia vine, *wu wei zi*, means 'five-flavoured fruit' as its red or pinkish-red fruit and seeds combine the tastes of sweet, sour, salty and bitter with spicy warmth. The species has a long history of use as a medicine in China and is included in the *Shen Nong Ben Cao Jing* (*The Drug Treatise of the Divine Countryman*), the earliest extant Chinese pharmacopoeia (see page 16). The fruit are also eaten (usually dried) as a food, or are incorporated into fruit juices and used to make jam, pastes and sweets. They are considered to be a nutraceutical, that is a food or dietary supplement considered to have some beneficial effect on health.

There are around 22 species in the genus *Schisandra*, most of which are native to Southeast Asia. One species, the star vine (*Schisandra glabra*), is found in North America, a disjunct distribution that is a remnant of a flora that stretched across the northern hemisphere before temperatures cooled 5 million to 2 million years ago.

ABOVE Dried fruit of Chinese magnolia vine (*Schisandra chinensis*) are used as a food, medicine and a nutraceutical.

AFRICA

Africa is the second-largest continent and stretches past the tropic of Cancer in the north and the tropic of Capricorn in the south. It encompasses a wide range of habitats, from deserts to rainforests, and from coasts to high mountain ranges. This diversity is reflected in the continent's flora and in its people, their cultures and their systems of medicine. Knowledge and traditions have largely been passed down by oral routes, so although African medicine is probably the world's oldest, it is among the least well known.

ABOVE **Rooibos or redbush (*Aspalanthus linearis*) has a restricted distribution, being found only in the fynbos vegetation of the Cederburg mountains in South Africa's Western Cape province.**

Stimulating substances

A common thread through African traditional medicine is a recognition that both psychological and physical aspects of an illness must be addressed. The use of stimulants and hallucinogenic plants can play an important role in determining the cause of an illness, and also in cultural traditions. In regions of west Central Africa, a drink made from the bark of yohimbe (*Corynanthe johimbe*; see page 196) in the coffee family (Rubiaceae) is used as a stimulant in ceremonies, and it is also drunk during initiation rituals, such as those of young Masai warriors in East Africa. Some tribes combine yohimbe with the roots of iboga (*Tabernanthe iboga*; see page 69) in the dogbane family (Apocynaceae). The hallucinogenic properties of iboga are also an integral part of the Bwiti religion, including in the diagnosis of illness. This religion is practised by a few tribes in Gabon, Equatorial Guinea and Cameroon, and in some areas Christianity has been incorporated into the more traditional ancestor worship and other beliefs.

Other African plants have had an impact around the world. The seeds of three species of coffee (*Coffea* spp.; see page 66) in the coffee family have become an integral part of many people's daily routine due to the presence of the mild stimulant alkaloid caffeine. A less widely used caffeine-free tea is made from the leaves and stems of rooibos or redbush (*Aspalanthus linearis*), a shrub in the legume family. Rooibos has been used traditionally in South Africa for asthma, headaches, nausea and depression.

Quite an ordeal

Until the second half of the nineteenth century, the innocence or guilt of someone accused of a crime was often determined using 'trial by ordeal'. During such a trial, the accused would eat or drink a poisonous substance and if they survived they were considered innocent, whereas death would point to their guilt. The Calabar bean (*Physostigma venenosum*; see page 72), a member of the legume family, was an ordeal poison used in Nigeria, while the Somali arrow poison (*Acokanthera*

schimperi) and kombe (*Strophanthus kombe*), both in the dogbane family, were used for ordeals and also as arrow poisons for hunting (see page 30). The highly toxic compounds in these plants have provided useful pharmaceutical drugs.

Poison bulb (*Boophone disticha*) in the amaryllis family (Amaryllidaceae) was used as an arrow poison in parts of southern Africa. The alkaloids present in the bulb would disorient the prey, while compounds from other plants or snake venom, with which the bulb extracts were combined, would kill them. Poison bulb is still used by healers to induce trances, either in themselves or in the patient during diagnosis. The bulb was also traditionally used externally on wounds, and internally as an emetic, sedative and analgesic – although this has resulted in serious cases of poisoning.

LEFT **Flowering and fruiting branches of Somali arrow poison (*Acokanthera schimperi*). The Greek philosopher Theophrastus may have been referring to this plant when he described roots used for arrow poisons in Ethiopia in the third century BCE.**

GOING BANANAS

Around a hundred of the continent's most important medicinal plants were included in the African Pharmacopoeia, published in 1985–1986, with descriptions and methods for how to prepare and analyze them. More recently (2013), an African Herbal Pharmacopoeia covering more than 50 plants has been produced. Plants used traditionally in Africa and also currently as herbal preparations in other parts of the world include devil's claw (*Harpagophytum procumbens*; see page 146) in the sesame family (Pedaliaceae). Other plants are valued for their reputed medicinal properties and their role as part of the diet, including the Ethiopian banana (*Ensete ventricosum*) in the banana family (Musaceae). The pseudostems and underground organs (corms) of this plant are a staple food for more than 20 million people in Ethiopia, but different landraces are also used as traditional medicines. New plant species are being discovered annually that could hold the key for future medicines discovery (see page 67), so long as this can be done in a way that protects the biodiversity of the continent.

LEFT **Ethiopian banana (*Ensete ventricosum*) is grown in gardens, including in temperate regions, where its striking leaves give a tropical effect. The prepared corms and pseudostems of certain cultivars are traditionally taken internally for wounds and for setting broken bones.**

Go with the flow

An enlarged prostate, known as benign prostatic hyperplasia (BPH), affects most men as they age, with symptoms that include increased frequency and difficulty in passing urine. Conventional drugs and surgery are available for BPH, depending on the stage and cause. For earlier management of symptoms, herbal preparations and dietary supplements are a popular, but less proven, approach.

THE PROSTATE PALM

PLANT:
Serenoa repens (W.Bartram) Small (syn. *Serenoa serrulata* (Michx.) Hook.f.)
COMMON NAME(S):
saw palmetto, sabal palm
FAMILY:
palm (Arecaceae)
ACTIVE COMPOUND(S):
NATURALLY OCCURRING:
phytosterols (*beta*-sitosterol);

fatty acids (including lauric and oleic acids)
MEDICINAL USES:
MAIN: benign prostatic hyperplasia
OTHER: urinary tract infections
PARTS USED:
fruit

Beta-sitosterol

ABOVE **The phytosterol *beta*-sitosterol may contribute to the activity of saw palmetto (*Serenoa repens*) and African plum (*Prunus africana*) in benign prostatic hyperplasia.**

For a small palm with saw-like teeth along the margins of its leaf stems, saw palmetto is a fitting name, *palmetto* being Spanish for 'small palm'. Saw palmetto is the only species in the palm genus *Serenoa* and is native to coastal plains extending from Louisiana eastwards to South Carolina. Native Americans from the region that is now Florida ate saw palmetto fruit (see box), and European settlers adopted the plant for medicinal uses in cases of infertility, impotence, digestive problems and urinary tract infections, and as a tonic and sedative. It is, however, for the relief of symptoms of an enlarged prostate in men that it has become most widely used. Today, saw palmetto is one of the top-selling herbal supplements in both North America and Europe. It is usually taken in the form of an extract made from the partially dried, ripe fruit, including the seed.

The accumulation of dihydrotestosterone (DHT) is considered to contribute to BPH. Laboratory studies have shown that saw palmetto extracts prevent binding of DHT at androgen receptors, and it inhibits the enzyme 5-*alpha* reductase, which converts testosterone to DHT. These properties are thought to be due to the presence of phytosterols such as *beta*-sitosterol, although other constituents might also play a role. While these studies did indicate the potential for

RIGHT **The upright stems of the saw palmetto (*Serenoa repens*) can grow to 2 m (6.5 ft) tall and bear fan-shaped leaves that reach 1 m (3 ft) in diameter. The stems can also grow along or under the ground, making the plant resistant to fire, which is a natural occurrence in the habitats in which it grows.**

Rotten cheese

The taste of saw palmetto fruit was not to the liking of the first Europeans who tried it, with one describing it as 'rotten cheese steeped in tobacco'. The Native Americans were not all that keen on the taste either and didn't eat the fruit in quantity. One trick was to make a drink from the juice and then add plenty of sugar. Despite this drawback, the fruit were an important part of the diet as they are rich in fatty acids and digestible carbohydrates. European settlers also fed the fruit to their animals – the cows that ate

them were said to produce more and richer milk than when given any other kind of food. Before livestock arrived in North America, saw palmetto plants were providing food or cover for more than 100 bird, 27 mammal, 25 amphibian and 61 reptile species. These included the American black bear (*Ursus americanus*), which often gave birth in the protective cover of saw palmetto thickets.

LEFT **Dried saw palmetto (*Serenoa repens*) fruit.**

saw palmetto to be effective at slowing prostate enlargement and relieving symptoms of BPH, the numerous clinical trials that have been undertaken have shown mixed results. Nevertheless, positive effects have been reported with some trials, and the popularity of saw palmetto seems secure.

PRUNES AND NETTLES

Two other plants have shown potential for improving urinary tract symptoms in men with BPH, and are often used in preparations with saw palmetto. African plum or prune, also known as pygeum (*Prunus africana*, syn. *Pygeum africanum*), is a small evergreen tree in the rose family (Rosaceae) from southern and Central Africa. As with the saw palmetto, the bark of African plum contains phytosterols such as *beta*-sitosterol, which are at least partly responsible for its action. Because the bark is the medicinal part of the plant, harvesting can be very damaging and there are concerns over its sustainability.

There are no such concerns for the second plant, the stinging nettle (*Urtica dioica*) in the nettle family (Urticaceae), which is native to Eurasia and now widely distributed around the world. Its roots are used as a herbal remedy for the symptomatic relief of BPH, and although they do contain small amounts of phytosterols, it is thought that their activity is due to different compounds, including lignans.

RIGHT **Trade in African plum (*Prunus africana*) across international borders has been regulated by the Convention on International Trade in Endangered Species of Wild Flora and Fauna (CITES) since 2017.**

Pass the sugar

Diabetes mellitus, literally 'fountain honey', was recognized at least 2,000 years ago as a disease that caused people to pass large amounts of sweet-tasting urine. One of the first-line pharmaceutical drugs used to treat diabetes today, metformin, has its origins in an alkaloid produced by goat's rue. With diabetes affecting an increasing number of people worldwide, other plants – including bitter gourd – offer a rich source of potential leads for the development of new therapeutic approaches.

GOAT'S RUE

PLANT:
Galega officinalis L.
COMMON NAME(S):
goat's rue, French lilac
FAMILY:
legume (Fabaceae)
ACTIVE COMPOUND(S):
NATURALLY OCCURRING:
alkaloid (galegine)

SYNTHESIZED FROM NATURAL LEAD: metformin
MEDICINAL USES:
MAIN: to lower blood glucose
OTHER: to stimulate milk flow
PARTS USED:
flowering stems

Galegine   Metformin

ABOVE **Goat's rue (*Galega officinalis*) contains the alkaloid galegine, which was used to develop the drug metformin, a pharmaceutical for type II diabetes.**

Goat's rue is a perennial herb that grows to 2 m (6 ft) tall and has a distribution extending from central and southern Europe east to Pakistan. Traditionally, it was fed to cattle to increase milk yields, and this practice was recognized in the genus name *Galega*, derived from the Greek word meaning 'milk'. One of goat's rue's traditional medicinal uses was the control of diabetes (see box). The active compound, isolated in the mid-1800s, was found to be a guanidine derivative that was given the name galegine. It wasn't until 1918–1920 that guanidine and, shortly after, galegine and similar compounds were found to lower blood glucose in laboratory studies. They were used to treat diabetes for a time but were abandoned in the 1930s due to their short duration of activity and their toxicity, as well as the introduction of insulin as a medication.

LUCK AND JUDGEMENT

Biguanides were initially synthesized by the fusion of two guanidine units in 1879, and a biguanide called metformin (dimethylbiguanide) was synthesized in 1922. Initial studies of metformin and other biguanides, however, revealed that high doses were required to

LEFT **Goat's rue (*Galega officinalis*) grows in moist grassland and disturbed ground. It has naturalized outside its native range and is considered to be a noxious weed as it is poisonous, particularly to sheep, and can be invasive.**

achieve modest reductions in blood glucose, and the compounds were not studied further. By coincidence, in the 1940s a biguanide-based anti-malarial agent was developed and its structure later modified to create metformin in the search for further such agents. When metformin was tested as an anti-malarial in the Philippines in 1949, it was found to be effective against influenza, giving rise to the anti-influenza agent flumamine.

One effect in some influenza patients treated with metformin (flumamine) was a reduction in blood glucose levels. This was investigated in 1956 by a French physician, Jean Sterne (1909–1997), who had been involved with a disappointing study of galegine many years earlier, and through his work metformin was developed as a drug for diabetes. Other biguanides were also investigated; of these, phenformin and buformin were found to lower blood glucose. However, unlike metformin, their use as therapeutic agents was relatively short lived due to their association with life-threatening lactic acidosis. Metformin's anti-diabetic activity has more than one mechanism. It sensitizes cells to insulin, resulting in greater uptake of glucose by muscle and adipose tissue, and it also supresses the production of glucose by the liver. Metformin is additionally being investigated for any potential role in cancer therapy.

LEFT Goat's rue (*Galega officinalis*) leaves are impari-pinnate, being formed of opposite pairs of leaflets and a terminal leaflet. It bears upright racemes of lilac or white flowers.

Noxious weed

One of the earliest written records of the use of goat's rue to treat symptoms of diabetes, namely thirst and frequent urination, was in Volume XXI of *The Vegetable System*, published in 1772 by the English botanist John Hill (*c.* 1714–1775). In medieval Europe and later, it was used for a variety of conditions, including worms, fever, pestilence and 'falling sickness' (epilepsy), and applied externally for skin ulcers. It was also traditionally used by nursing mothers to stimulate the flow of milk, and herbal preparations for this purpose were available in Europe until recently. Their use is no longer recommended, however, as the health risks are considered to outweigh any benefits.

The toxicity of goat's rue to domestic animals became apparent when it was introduced to the United States in 1891 as a potential fodder crop. In feeding studies, it was found to be unpalatable and poisonous to cattle, sheep and horses. Unfortunately, it escaped from the abandoned research plots and is now a noxious weed requiring control in a number of states.

LEFT The English botanist John Hill used goat's rue (*Galega officinalis*) to treat symptoms of diabetes.

BITTER GOURD

PLANT:

Momordica charantia L.

COMMON NAME(S):

bitter gourd, bitter melon

FAMILY:

cucumber (Cucurbitaceae)

ACTIVE COMPOUND(S):

NATURALLY OCCURRING:

cucurbitane-type

triterpenoids; steroidal saponins (charantin)

MEDICINAL USES:

MAIN: type 2 diabetes

OTHER: skin infections

PARTS USED:

leaves, fruit

The scientific name for the genus *Momordica* is derived from the Latin word meaning 'to bite' and could refer to the bitterness of the fruit. Nevertheless, the fruit of many species are eaten as a vegetable when mature but still green, and numerous local cultivars exist. The bitter gourd is one of the most widely distributed members of the genus, being native to tropical and subtropical Africa, India, Bangladesh and Laos, and naturalized in much of South and Central America and elsewhere. This annual climbing vine can reach 5 m (16 ft) in height and bears fruit that resemble a warty cucumber.

In regions where bitter gourd is native or naturalized, various parts of the plant are used medicinally for conditions including anaemia, fever, gout, rheumatism and urinary stones. The fruit and leaves have been used traditionally for diabetes and some laboratory studies have looked at their mechanisms and effectiveness. Both of these plant parts contain cucurbitane-type triterpenoids that reduce insulin resistance, and a mixture of steroidal saponins called charantin that also show an ability to lower blood glucose levels. One recent small clinical trial suggested that dried bitter gourd fruit, with the seeds removed (use of these has been associated with adverse effects), could reduce some long-term complications that arise from raised blood glucose levels. Despite the widespread traditional use of bitter gourd leaves and fruit, and anecdotal reports of beneficial effects, clinical evidence to support their use in people with diabetes is lacking.

PREVENTING POST-PRANDIAL SPIKES

Iminosugars mimic carbohydrates, but the oxygen atom of the sugar ring is replaced by a nitrogen atom. Although the iminosugar 1-deoxynojirimycin (1-DNJ) was initially synthesized chemically, it was later found to occur naturally in some plants and bacteria. Two sources are the Asiatic dayflower (*Commelina communis*; see page 103), which is used in traditional Chinese medicine, and white mulberry (*Morus alba*) in the

1-Deoxynojirimycin (1-DNJ)

ABOVE White mulberry (*Morus alba*) leaves are rich in iminosugars, including 1-deoxynojirimycin, which was used as a template to design new drugs for diabetes.

LEFT The bitter gourd (*Momordica charantia*) vine bears warty fruit that are used as a vegetable and medicinally when still green.

Novel sources of insulin

Insulin-dependent diabetes (type 1 diabetes) occurs when the pancreas no longer produces sufficient or indeed any of the peptide hormone insulin. The role of the pancreas in diabetes was suggested in 1889 by two German scientists, the physiologist and pathologist Oskar Minkowski (1858–1931) and the physician Joseph von Mering (1849–1908), who found they could trigger a condition similar to diabetes by removing the pancreas from a dog. It was in the early 1900s that insulin was isolated and shown to be the active hormone. Humans were first treated for diabetes with pig and cow insulin, but as these differ from human insulin it could be difficult to manage the timing and dose. Since the 1980s, the main treatment has been human insulin manufactured by genetically modified yeast and bacteria, but this is expensive to produce. More recently, genetic modification undertaken on safflower (*Carthamus tinctorius*) in the daisy family created plants that manufactured pro-insulin (the precursor to human insulin) in their 'seeds' (achenes). It was hoped that this would provide a cheaper source of insulin. Although initial tests showed that the insulin from processed seeds, dubbed 'prairie insulin', was identical to human insulin and safe to use, the pioneering work was abandoned in 2012.

LEFT Safflower (*Carthamus tinctorius*) is used medicinally and its florets are a cheap alternative to saffron (see page 61).

mulberry family (Moraceae), used in traditional Chinese, Japanese, Korean and African systems of medicine. Traditional applications of these plants include the control of blood sugar levels. The drug miglitol is a close analogue of 1-DNJ and is approved by the United States Food and Drugs Administration as an anti-diabetic. Like 1-DNJ, it is an *alpha*-glucosidase inhibitor, and when taken at the start of a meal it prevents the digestion of complex carbohydrates and their breakdown into glucose.

BELOW The rind of the ripe orange fruits and the red seeds of bitter gourd (*Momordica charantia*) have a bitter taste and are toxic if eaten.

CHAPTER 6

COUGHS AND SNEEZES

Symptoms associated with colds and other
infections of the respiratory system can
be eased and treated by a variety of plant
compounds. Some of them clear congestion in
the nose or on the chest, and in addition can
be useful for asthma. Viruses and other
microorganisms can be tackled by compounds
that support the immune system, or fight the
infection itself, including the influenza and
human immunodeficiency viruses and the
parasite that causes malaria.

Coneflower (*Echinacea* sp.)

NATURAL COMPOUNDS FOR FIGHTING INFECTIONS

The moist, warm, nutrient-rich environment of the human body provides ideal conditions for numerous microorganisms to live and multiply. A complex immune system and automatic physiological responses, such as a raised body temperature, have evolved that target any invaders and make the environment less hospitable. In this chapter, we look at plants that help us fight infections and treat some of the symptoms.

ABOVE **The presence of ephedrine-type alkaloids in Mormon tea (*Ephedra nevadensis*) is in question. It is one of the North American species of ephedra, being native to Oregon and southwards to northwest Mexico.**

TAKE A DEEP BREATH

Colds and the flu are caused by viral infections of the upper respiratory tract. While a few drugs can target the virus responsible, treatment is usually limited to easing the symptoms while the body's own immune system fights off the attack. Ephedrine and pseudoephedrine from Chinese ephedra (*Ephedra sinica*; see page 120) in the ephedra family (Ephedraceae) have been added to cold-relief preparations as they can reduce nasal congestion, but misuse has led to tighter controls. Congestion of the nasal cavities can also be loosened by inhaling essential oils from plants such as species of eucalyptus (*Eucalyptus* spp.; see page 121) in the myrtle family (Myrtaceae). Infection can cause the lungs to become congested, and compounds that thin the mucus so that it can be more easily expelled by coughing are called expectorants. Guaifenesin is an expectorant that is synthesized from guaiacol, a compound found in several plants, including roughbark lignum vitae or guaiac (*Guaiacum officinale*) and holywood lignum vitae (*Guaiacum sanctum*), two trees in the caltrop family (Zygophyllaceae; see page 122). Compounds that relax the airways can be useful during colds, but several also have found applications for asthma. Among these is khellin from khella (*Visnaga daucoides*; see page 124) in the carrot family (Apiaceae). In addition to being used as a remedy itself, khellin is the starting point for the synthesis of sodium cromoglicate, a drug used to manage asthma and other types of allergy.

IMMUNITY ASSISTANCE

Boosting our immune system has the potential to shorten the length of an infection, reduce its severity, or even stop the cold or seasonal flu before it gets a hold. The daisy family (Asteraceae) brings us the most widely used immune system stimulant, echinacea, from purple coneflower (*Echinacea purpurea*; see page 126) in

ABOVE **Malabar nut or vasaka (*Justicia adhatoda*, syn. *Adhatoda vasica*) in the acanthus family (Acanthaceae) is from India and eastwards to Laos. It contains a quinazoline alkaloid, peganine (vasicine), which acts as a bronchodilator and from which bromhexine and ambroxol are synthesized.**

particular. When a new strain of the influenza virus evolves for which humans have limited immunity, a pandemic can occur that requires more than a helping hand for the immune system or the treatment of symptoms. Anti-influenza therapies include oseltamivir, which can be synthesized using shikimic acid as the starting material. This compound is usually extracted from species of star anise (*Illicium* spp.), including the culinary spice star anise (*Illicium verum*; see page 130) in the magnolia vine family (Schisandraceae).

INSPIRED BY TRADITION

The human immunodeficiency virus (HIV) reduces the body's ability to respond to infections and diseases because it attacks our immune system. A damaged immune system can result in potentially life-threatening illnesses that are known as acquired immune deficiency syndrome (AIDS). Aggressive therapies can slow the progression of HIV but they are expensive and do not yet provide a cure. Plant compounds with different mechanisms of action are under investigation, including prostratin. This was identified as the compound responsible for the traditional use of mamala (*Homalanthus nutans*; see

page 132) in the spurge family (Euphorbiaceae) to treat a different viral infection, hepatitis.

Viruses are not the only biological agents that can attack the human body. Malaria is caused by four species of single-celled parasites in the genus *Plasmodium* that are transmitted to humans by female *Anopheles* mosquitoes. The use of traditional plant remedies led to the discovery of two important treatments. For several centuries, the bark of a few species of the cinchona tree (*Cinchona* spp.; see page 135) in the coffee family (Rubiaceae) was the main prevention and treatment for malaria. Artemisinin and semi-synthetic derivatives from sweet wormwood (*Artemisia annua*; see page 134) in the daisy family are now part of the response to these life-threatening parasites.

Cutting congestion

A stuffy or blocked nose – a feature of many colds – is caused by congestion of the nasal cavities and compounds that can alleviate it are therefore called decongestants. They include some found in particular species of ephedra (*Ephredra* spp.), which are available in pharmaceutical cold remedies but are also abused. Many other plants offer relief based on their traditional use as herbal preparations or inhalants.

CHINESE EPHEDRA

PLANT:

Ephedra sinica Stapf and other species

COMMON NAME(S):

Chinese ephedra, desert tea

FAMILY:

ephedra (Ephedraceae)

ACTIVE COMPOUND(S):

NATURALLY OCCURRING:

alkaloids (ephedrine, pseudoephedrine, norephedrine)

MEDICINAL USE(S):

MAIN: decongestant

OTHER: bronchodilator for asthma

PARTS USED:

stems, leaves

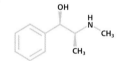

Ephedrine

Ephedra has been used in traditional Chinese medicine for thousands of years, with the earliest written record of it found in the *Shen Nong Ben Cao Jing* (*The Drug Treatise of the Divine Countryman*) dating to *c.* 25–200 CE but incorporating a much older text (see page 62). Many species of ephedra are used medicinally, including for a number of respiratory complaints, but it is the stems of Chinese ephedra, one of the three species known as *ma huang* in traditional Chinese medicine, that are most widely employed.

Chinese ephedra contains several alkaloids, including ephedrine, which was first isolated in 1885 by the Japanese chemist Nagayoshi Nagai (1844–1929). Almost 30 years later, its stimulatory effect on the

LEFT **The amino alkaloid ephedrine occurs in Chinese ephedra (*Ephedra sinica*) and other *Ephedra* species. It is used pharmaceutically to relieve nasal congestion and has also been used to dilate the airways in people with asthma.**

sympathetic nervous system and similarity in action to the neurotransmitter adrenaline (epinephrine) were discovered and investigated. Ephedrine is now known to be more similar in action to another neurotransmitter, noradrenaline (norepinephrine). The ability of ephedrine to relax the muscles of the bronchial tubes within the lungs was found to be of benefit in the treatment of asthma, particularly as it can be taken by mouth – unlike adrenaline, which is effective only if injected. As demand increased, medical supplies of ephedrine were assured in the 1930s, first by an increase in the cultivation of ephedra in India and later by successful laboratory synthesis of the compound. By the 1950s, however, ephedrine was falling out of favour as a treatment for asthma, as alternative, safer treatments were developed (see page 124).

COMPROMISED COLD REMEDY

Ephedrine and pseudoephedrine function as a decongestant by reducing the swelling of blood vessels in the nose, and are therefore added to preparations for treating the symptoms of colds. Long-term use, however, can cause side effects such as anxiety, and the medications are addictive, resulting in symptoms of withdrawal if stopped abruptly. Ephedrine-type alkaloids are also subject to abuse by people wishing to exploit them as a reputed slimming aid or to increase muscular performance. In fact, they are now banned substances in many sports; the Argentinian footballer Diego Maradona (b. 1960) is among those who have tested positive. Such abuse can also have serious health consequences, including hyperthermia, heart failure and death. In addition, ephedrine is structurally similar to the illegal drug methamphetamine, known as meth or crystal meth. Small- and large-scale illicit laboratories sprang up, particularly in the United States, to manufacture methamphetamine from ephedrine obtained from cold preparations bought over the counter, and from imported ephedra herbal material. Ephedra herb is itself abused as a 'herbal high'. For all these reasons, the use of ephedrine-type alkaloids in cold preparations and of herbal ephedra is now regulated in many countries.

LEFT Chinese ephedra (*Ephedra sinica*) is native to north and northeast China, Mongolia and southeast Russia. Most of the plants in this book are flowering plants but, like yew (*Taxus* spp.; see page 206), Chinese ephedra is a gymnosperm. The rigid stems of this shrub, with their scale-like leaves, are harvested in the autumn.

Herbal inhalants

One effective method of reducing nasal congestion is through breathing in essential oils and aromatic compounds from plants. These can be delivered in chest rubs and nasal sprays, or by steam inhalation and baths. Plants most commonly used in this way include species of eucalyptus, such as the blue gum tree (*Eucalyptus globulus*) from southeast Australia and Tasmania, in which the main volatile constituent of the leaves is 1,8-cineole. The aromatic camphor from the camphor tree (*Cinnamomum camphora*) in the laurel family (Lauraceae) can be another ingredient of inhaled decongestants (see also page 145).

ABOVE The blue gum tree (*Eucalyptus globulus*) can grow to 60 m (195 ft) tall and bears white flowers. Its leaves and the essential oil that is distilled from them are widely used medicinally for their antiseptic and expectorant properties.

Get it off your chest

Coughs are an annoying and potentially painful symptom of many colds and chest infections. Herbal preparations may soothe a dry or tickly cough, or help to loosen phlegm. Here, we look at the source of a compound that is an effective drug for a productive chesty cough.

LIGNUM VITAE

PLANT:
Guaiacum officinale L. and *Guaiacum sanctum* L.

COMMON NAME(S):
roughbark lignum vitae, guaiac; holywood lignum vitae, holywood

FAMILY:
caltrop (Zygophyllaceae)

ACTIVE COMPOUND(S):

NATURALLY OCCURRING:
phenolic (guaiacol); volatile oil (guaiol)

SEMI-SYNTHESIZED:
guaifenesin

MEDICINAL USES:

MAIN: expectorant

OTHER: syphilis, externally for rheumatism

PARTS USED:
wood, resin

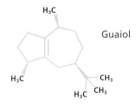

Guaiol

LEFT **Guaiol is a sesquiterpene constituent of the volatile oil from lignum vitae (*Guaiacum officinale* and *G. sanctum*) and has been used as an insecticide.**

Lignum vitae, literally 'wood of life', is a valuable wood sourced from two slow-growing tree species, although roughbark lignum vitae (also called guaiac) is considered to be the more important. Both species are native to Florida, the Caribbean and Panama, with roughbark lignum vitae also found in northern South America; holywood lignum vitae has a more northerly distribution through Central America and southern Mexico. With their striking blue flowers, the two species are national emblems of Jamaica and the Bahamas. The wood contains a resin that is used medicinally, either in the form of wood shavings or once it has been extracted from the wood by heating.

COUGH IT UP

Traditionally, lignum vitae was used for sore throats, tonsillitis and upper respiratory tract infections (URTIs). Guaifenesin (guaiacol glyceryl ether) is a semi-synthetic derivative of guaiacol that is found in a number of plants (see also page 96), but is usually obtained from guaiac resin. The drug was approved by the United States Food and Drug Administration in 1952 and is now classed as an expectorant. It thins mucus that may be congesting the airways, making it easier to cough it up. Due to the presence of guaiol, another volatile oil constituent, guaiac resin, is used in test strips to detect blood in faeces. The haem moiety from haemoglobin helps hydrogen peroxide oxidize the paper-embedded guaiac, converting guaiol into a blue-coloured quinone product called guaiazulene.

LEFT *Hyacum et Lues Venerea* (c. 1570), showing syphilis being treated with a drug prepared from the resin of lignum vitae (*Guaiacum* sp.). The preparation of the drug is seen at right, while the contraction of the disease is shown in the panel on the wall at centre. Engraving by Philip Galle (1537–1612), after the original painting by Johannes Stradanus (1523–1605).

RIGHT **Flowering branch of roughbark lignum vitae (*Guaiacum officinale*).**

Pox and propellers

When Europeans first encountered lignum vitae in the West Indies in the late 1400s, it was already being used medicinally for a range of conditions. Syphilis, known as the 'pox', was added to the list after the explorers introduced the disease to the islands. Lignum vitae soon became an ineffectual syphilis 'cure' in Europe, the alternative being mercury. German scholar Ulrich von Hutten (1488–1523) described his treatment with guaiacum in 1519; he died four years later. Today, the anti-inflammatory properties of lignum vitae are valued in herbal preparations applied externally to improve circulation and relieve painful muscles and joints, particularly in rheumatism and rheumatic arthritis.

The resin in lignum vitae gives the wood unusual properties. It is extremely dense and sinks in water, and is also self-lubricating. This makes it ideal for making propeller shafts and ball bearings. So much of the wood was exported to Europe by early merchants that, in 1701, a law was passed in Martinique to protect the trees. Today, habitat destruction is thought to be the most serious threat and trade is regulated under the Convention on International Trade in Endangered Species of Wild Flora and Fauna.

Short of breath

Asthma attacks are characterized by a difficulty in breathing, with the frequency and duration of episodes varying between individuals. Worldwide, the condition is thought to affect 339 million people and causes an estimated 1,000 fatalities a day. Asthma cannot be cured, but khella was the starting point for a prophylactic drug and many other plants have also provided a number of useful compounds to treat the symptoms.

ABOVE **Flowering khella (*Visnaga daucoides*) plants.**

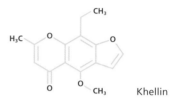

ABOVE **Khella (*Visnaga daucoides*) contains khellin, which can dilate the airways and has been used to alleviate asthma.**

ASTHMA AND ALLERGY

PLANT:

Visnaga daucoides Gaertn. (syn. *Ammi visnaga* (L.) Lam.)

COMMON NAME(S):

khella, toothpick weed

FAMILY:

carrot (Apiaceae)

ACTIVE COMPOUND(S):

NATURALLY OCCURRING:

furanochromones (khellin, visnagin); pyranocoumarins (visnadin)

SYNTHESIZED FROM NATURAL LEAD: sodium cromoglicate [cromolyn sodium]; also amiodarone for arrhythmias

MEDICINAL USES:

MAIN: asthma, hay fever, allergic rhinitis

OTHER: angina pectoris

PARTS USED:

fruit

Khella is an annual or biennial herbaceous plant that is native to the Mediterranean, Ethiopia, Eritrea and Iran. It has the typical umbel-shaped flower heads of members of the carrot family. The fruit have been used medicinally in the Middle East and elsewhere for the relief of respiratory tract complaints, such as in asthma, and for angina pectoris and kidney conditions. These traditional uses have inspired research into the active compounds and their medical applications. For example, the furanochromone khellin was the lead for amiodarone, an anti-arrhythmia drug (see page 41).

Khellin and other compounds in khella, such as visnagin, have been found to induce relaxation of smooth muscles (spasmolytic) by inhibiting transport of calcium through the membrane of their cells. In 1947, khellin administered in the form an intramuscular injection was reported to provide relief for bronchial asthma and was adopted in several countries. There were, however, side effects from long-term use, including nausea and vomiting.

GUINEA PIG SCIENTIST

In the early 1950s, a team in the United Kingdom started to look for derivatives of khellin that were effective when inhaled or taken by mouth. Roger Altounyan (1922–1987), an Anglo-Armenian doctor and pharmacologist, was part of that team. As an asthma sufferer, he took the rather unusual approach of inducing asthma attacks in himself – probably around 1,000 times – in order to test potential treatments. Sodium cromoglicate was synthesized in 1964 and tested the following year. Applied in the form of an aerosol, it is now used as a preventative (prophylactic) treatment for allergic asthma; it is also administered in various other forms, including as eye drops for allergies. Sodium cromoglicate has a different mechanism of action than khellin, as it stabilizes mast cells in the lungs and prevents the allergen-induced release of the substance that causes constriction of the airways during an asthma attack.

ABOVE One of the common names for khella (*Visnaga daucoides*), toothpick weed, derives from a traditional North African use of the dried rays of the fruit head.

Easing breathing

In addition to khellin, other plant compounds that relax the airways have been found to be effective in the treatment of asthma. They include compounds discussed elsewhere, such as caffeine and theophylline from plants such as arabica coffee (*Coffea arabica*; see page 67) in the coffee family, and the ephedrine-type alkaloids from species of ephedra (see page 120). The opium poppy (*Papaver somniferum*; see page 140) in the poppy family (Papaveraceae) is best known for its painkilling compounds, but it also produces papaverine and narcotine, which are spasmolytic in action. Finally, corticosteroids are probably the most important drugs for asthma, with diosgenin from certain species of yam (*Dioscorea* spp.; see page 186) in the yam family (Dioscoreaceae) and hecogenin in sisal (*Agave sisalana*; see page 162) in the asparagus family (Asparagaceae) being sources of corticosteroid precursors.

ABOVE *Dioscorea deltoidea*, a species of yam, is native to the Himalayas, through China to Myanmar. It is one of the richest sources of diosgenin but exploitation means that it is now endangered in some parts of its natural range.

Boosting immunity

When a cold or flu virus attacks the body, the immune system mobilizes to fight it off. Herbal preparations may have a stimulating effect on the immune system, with the potential to reduce the duration and severity of infections. Echinacea is probably the best known and studied, but others include elder (*Sambucus nigra*) and andrographis or kalmegh (*Andrographis paniculata*).

PRAIRIE PROMISE

PLANT:
Echinacea purpurea (L.) Moench

COMMON NAME(S):
purple coneflower, echinacea

FAMILY:
daisy (Asteraceae)

ACTIVE COMPOUND(S):

NATURALLY OCCURRING:
alkylamides; caffeic acid derivatives

MEDICINAL USES:

MAIN: immunostimulant

OTHER: externally for minor skin conditions

PARTS USED:
aerial parts, roots

Most people will suffer from a cold at some time during the year, so it may be no surprise that echinacea is one of the most commonly used herbal remedies in Europe and North America. Echinacea preparations can consist of one or more of three species. Purple coneflower has been well studied and is now widely cultivated for use in such preparations, but pale coneflower (*Echinacea pallida*) and narrow-leaf coneflower (*E. angustifolia*) are also used. (Some classifications of the genus treat narrow-leaf coneflower as a variety of the pale coneflower with the scientific name *E. pallida* var. *angustifolia*).

Echinacea has probably been used medicinally for thousands of years by the native people of North America to treat both external and internal conditions,

including wounds and burns, snakebites, tonsillitis, toothache, colds and coughs. It was adopted by European settlers, who incorporated it into herbal products, and has been used in Europe since the early twentieth century. Clinical trials have largely looked at the use of preparations made from the roots or aerial parts of echinacea, particularly purple coneflower, in people with upper respiratory tract infections (URTIs) and colds. These trials have provided some evidence that it can reduce the severity and duration of URTIs, although other trials have shown no effects. Variation between the preparations being tested makes it difficult to draw conclusions. Studies have shown that echinacea can enhance the immune response, with a number of compounds, including alkylamides, contributing to the effects.

ELDER AND ANDROGRAPHIS

Elder is a large shrub or small tree in the moschatel family (Viburnaceae, syn. Adoxaceae) that is found through Europe to western Asia and widely naturalized elsewhere. Raw elderberries can be toxic, but syrups and vinegars made by heating dried berries may be used for colds and flu. There is some evidence that such remedies stimulate the immune response to viruses and prevent their attachment to the respiratory system, and also have the more direct effect of soothing a sore throat. Andrographis is a bitter-tasting annual herbaceous plant in the acanthus family (Acanthaceae) that is used in Ayurvedic medicine, where it is known as kalmegh, and also in other Asian systems of medicine. Clinical trials have shown that preparations made from andrographis leaves can alleviate the symptoms of URTIs. The active compounds are diterpene lactones, known as andrographolides, and studies found that they act by stimulating the immune system and through anti-inflammatory effects.

ABOVE Andrographis (*Andrographis paniculata*), which is native to India, is considered to be one of the most important and popular plants in traditional medicine in Asia, where it is used for a great variety of conditions.

LEFT The fruits and flowers of elder (*Sambucus nigra*) are used to make both alcoholic and non-alcoholic drinks, and the flowers are also a traditional remedy for hayfever.

LEFT Purple coneflower (*Echinacea purpurea*) can reach 1.2 m (4 ft) tall. It is native to North America, where it grows in prairies and open woodland, and sometimes in wetter ground near watercourses.

Hedgehog flowers

The German botanist Conrad Moench (Konrad Mönch; 1744–1805) included only one species, the purple coneflower, when he described the genus *Echinacea* in 1794. His choice of name from the Greek for 'hedgehog', *ekhinos*, alludes to the spiny centre to the 'flower' (flower head). Purple coneflower had earlier been described (in 1753) by the Swedish taxonomist Carl Linnaeus (1707–1778), who had placed it in the genus *Rudbeckia*, commonly known as the black-eyed Susans. The two genera share many characteristics and also the alternative common name of coneflower. They are also both native to central and eastern North America, and are widely grown in temperate gardens around the world. The colour of the ray florets can help to distinguish them: those of black-eyed Susans are yellow or orange, while those of echinaceas are usually purple, pink or white, although one species, the yellow coneflower (*Echinacea paradoxa*) can have yellow ray florets and has been widely used to breed cultivars.

COUGHS AND SNEEZES

OCEANIA

Oceania lies to the east of Malesia, with the island of New Guinea forming the boundary between the two regions. It includes Australia and the many Pacific islands of Melanesia, Micronesia and Polynesia. Of these, Australia has the most diverse range of habitats – from deserts to tropical rainforest, and from coastal fringes to mountains – and a flora of more than 30,000 plant species. Since the arrival of Europeans, pressure on land throughout the region for agriculture, logging and mining has threatened the survival of many species.

ABOVE **Pōhutukawa (*Metrosideros excelsa*) is also known as New Zealand Christmas tree as it flowers in December and January.**

Local traditions

Traditional medicine in Oceania has relied on local plants and ceremonies that were often administered by healers, with information passed down by word of mouth. Unlike in North America, very few local plants were adopted by European settlers and even fewer have come to international attention.

In Australia, the best known medicinal plants are eucalyptus (*Eucalyptus* spp.; see page 121) and tea tree (*Melaleuca alternifolia*; see page 170), both in the myrtle family (see also box). Much traditional knowledge has been lost as people were moved off their land and plants disappeared due to habitat loss. Surviving members of Aboriginal Australians and other indigenous communities have often adapted their practices to incorporate more easily available, non-native plants. Many have been reluctant to share their knowledge with outsiders, but some initiatives have sought to record the information for use by the community or for wider dissemination before it is lost.

Rongoā Māori is the traditional Māori healing system. It relies on New Zealand plants such as pōhutukawa (*Metrosideros excelsa*) in the myrtle family, whose inner bark is used to stop bleeding, for toothache and for diarrhoea; and harakeke (*Phormium tenax*) in the asphodel family (Asphodelaceae), whose roots and sap are used for wounds and intestinal worms.

In Samoa, plants used locally include ifiifi or makita (*Atuna racemosa* subsp. *racemosa*), in the cocoplum family (Chrysobalanaceae). An oil extracted from the cotyledons of mature ifiifi seeds is combined with the flowers of ylang ylang (*Cananga odorata*; see page 175) in the soursop family (Annonaceae), among other plant ingredients, and used for medical massages, particularly for pain in the back, neck or stomach. A study has shown that ifiifi oil may work by reducing inflammation due to the presence of COX-2 inhibitors.

Ceremonial peppers

Plants used during ceremonies include the hypnotic roots of kava (*Macropiper methysticum*; see page 56) in the

ANTIPODEAN ADVENTURES

The contribution of the English naturalist and botanist Sir Joseph Banks (1743–1820) to knowledge of the flora of Oceania is commemorated in the name of the genus *Banksia*, in the protea family (Proteaceae), which is native to Australia, Tasmania and New Guinea. Banks accompanied Captain James Cook (1728–1779) on the first voyage he took around the world on HMS *Endeavour* (1768–1771). Species of eucalyptus or gum trees are among those plants that Banks is credited with introducing to Europe from the region he visited. The Australian tea tree came to wider attention more recently.

Sugar cane (*Saccharum officinarum*), in the grass family (Poaceae), was first cultivated in New Guinea in about 6000 BCE. From there it was carried west by traders, reaching Europe around 700 CE, and then taken to the Caribbean by the explorer Christopher Columbus (*c.* 1451–1506) on his second voyage in 1493. Recent research has shown that sugar cane evolved in Polynesia. It is probably the most widely used plant native to Oceania as it is now the source of more than half the world's sugar.

ABOVE **A Papuan man chewing sugar cane (*Saccharum officinarum*) at a market in Mount Hagen, Western Highlands province, Papua New Guinea. On the day before harvesting, fields of sugar cane are often set on fire to remove the dead leaves, which makes sugar extraction easier.**

LEFT **Kawakawa (*Macropiper excelsum*, syn. *Piper excelsum*), native to New Zealand, is a small tree of coastal forests that reaches 5 m (16 ft) in height. Its broad leaves, as well as the fruit, have a peppery taste, and the male and female flowers are borne in separate, erect heads.**

pepper family (Piperaceae). Also in the genus is kawakawa (*M. excelsum*, syn. *Piper excelsum*), which is native to New Zealand, where the leaves are traditionally used by Māori as a flavouring in food and during a range of ceremonies, such as the opening of a new meeting house, the launch of a canoe and as a symbol of mourning. They have also been used to relieve conditions such as eczema and rheumatism, and as a general tonic.

The leaves of another member of the pepper family, betel vine (*P. betle*; see page 99), are used to wrap betel nuts, the seeds of the betel palm (*Areca catechu*) in the palm family (Arecaceae), in a quid that is chewed. Although both of these plants are native to Malesia and further west, they have been introduced to other parts of Oceania, where they are used socially, to suppress appetite and for the relief of fatigue.

Given the rich and unique flora of the Oceania region, it has promising potential as a source of important new plant compounds.

Influenza

Infection by the influenza virus – flu – can cause fever, a sore throat and runny nose, and exhaustion. There are several million severe cases of flu worldwide each year and up to half a million deaths. When pandemics occur, mortalities can reach 2 million, so treatments are held in reserve to tackle these outbreaks. A plant compound called shikimic acid is an important component in the manufacture of one of these.

STAR OF THE SHOW

PLANT:
Illicium verum Hook.f.

COMMON NAME(S):
star anise, Chinese star anise

FAMILY:
magnolia vine
(Schisandraceae)

ACTIVE COMPOUND(S):

NATURALLY OCCURRING:
shikimic acid

SEMI-SYNTHESIZED:
oseltamivir

MEDICINAL USES:

MAIN: influenza virus A
and B

OTHER: avian influenza virus
(bird flu)

PARTS USED:
fruit

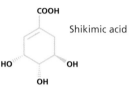

Shikimic acid

LEFT Shikimic acid occurs at high levels in star anise (*Illicium verum*) and is used as the starting material to synthesize the anti-influenza drug oseltamivir.

BELOW Star anise (*Illicium verum*) is thought to have originated in Guangxi province, China. Its fruit usually consist of around eight follicles, each of which splits open when ripe to reveal a single seed. Dried fruit are used medicinally, to flavour food and in perfume.

The star anise, which as its name suggests has an anise flavour, is used as a spice in Asian cuisine and also medicinally (see box). The flavour is due to the presence of phenlypropanoids, especially *trans*-anethole. Phenylpropanoids are one of a group of compounds that plants produce from shikimic acid. In fact, the shikimic acid (or shikimate) pathway produces numerous essential and secondary compounds in both plants and bacteria, but not animals. Shikimic acid is therefore found in many plants, but a few – including some species of the star anise genus (*Illicium* spp.) – contain significant amounts of this compound. Shikimic acid had been isolated from Japanese star anise (*Illicium anisatum*) – called *shikimi* in Japan – before its role in metabolism was recognized. When it was found that oseltamivir, a treatment for some types of influenza virus that is marketed under various names (including Tamiflu), could be manufactured most efficiently using shikimic acid as the starting material, it was star anise that chemists turned to as a readily available source.

STOPPING THE SPREAD

Virus particles consist of nucleic acid (DNA/RNA) surrounded by a protein matrix. The human immune system learns to recognize antigens, components of the surface of this matrix, and they also hold the key to treating a flu infection. Two surface antigens, called haemagglutinin and neuraminidase, are found on an influenza virus. Variations in the structures of these two antigens are used to identify different subtypes of flu – for example, H2N2, which is the designation given to the virus that caused the pandemic known as Asian

The right star

Star anise is used medicinally in many systems of medicine in Asia, and also in Europe, the United States and South America. The fruit are used as a digestive aid, and to loosen congestion of the respiratory tract (expectorant). In the early 2000s, several cases of poisoning from consumption of star anise tea were reported. They were traced to contamination by fruit of the Japanese star anise, which have a very similar appearance. Unlike star anise, Japanese star anise contains high levels of anisatin, a toxic sesquiterpene lactone that can cause seizures and other adverse effects. Laboratory analyses have now been put in place to detect such contamination.

flu in 1957. Haemagglutinin is a glycoprotein that can bind to receptors on human cells and then enables the viral nucleic acid to enter the cell. Neuraminidase is an enzyme that releases newly formed virus particles from infected cells and also stops the particles from sticking together.

The anti-influenza drug oseltamivir works by binding to the active site on neuraminidase, thereby inhibiting its action. Targeting neuraminidase therefore does not stop the initial infection of a cell, but it can stop the spread of the virus around the body. Because of this, oseltamivir is effective only if given shortly (up to two days) after flu symptoms develop, and it can also be used as a preventative treatment. Its use is, however, reserved for serious outbreaks to reduce the chances of the particular strain developing resistance to the drug and rendering it ineffective.

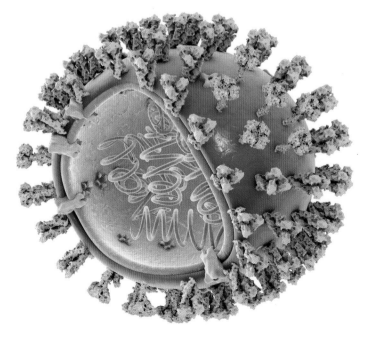

LEFT **Three-dimensional illustration of the influenza virus. This shows the two types of surface antigens (glycoprotein spikes) – haemagglutinin in blue and neuraminidase in green. Part of the virus envelope has been cut away to reveal the RNA fragments inside.**

HIV heroes

An estimated 37 million people are living with HIV/AIDS worldwide, with almost a million dying from the infection each year. Conventional approaches to HIV, known as antiretroviral therapy (ART), have been developed in response to such enormous figures, but they require lifelong medication. Plant compounds are offering new ways to attack the virus and offer hope for an eventual cure.

SOUTH PACIFIC SAVIOUR

PLANT:
Homalanthus nutans
(G.Forst.) Guill.

COMMON NAME(S):
mamala

FAMILY:
spurge (Euphorbiaceae)

ACTIVE COMPOUND(S):

NATURALLY OCCURRING:
prostratin

SYNTHESIZED FROM NATURAL
LEAD: prostratin analogues

MEDICINAL USES:

MAIN: intestinal complaints

OTHER: hepatitis virus

PARTS USED:
stem bark and wood

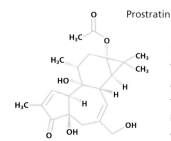

Prostratin

LEFT **Prostratin is a diterpene found in mamala (***Homalanthus nutans***) and has been studied for its effects against the human immunodeficiency virus (HIV).**

ABOVE **Mamala (*Homalanthus nutans*) tree branch.**

RIGHT **Coloured scanning electron micrograph of a cell infected with the human immunodeficiency virus (HIV). The small spherical virus particles, visible on the surface as yellow dots, are in the process of budding from the cell membrane.**

In the 1980s, the American ethnobotanist Paul Alan Cox (b. 1953) worked with a community in the Falealupo rainforest in Samoa, where he collected plants and recorded the traditional uses described to him by local healers. One plant was of particular interest: the mamala tree, which is native to islands of the South Pacific Ocean. A water infusion of the stem bark and wood of the mamala tree was used for intestinal complaints and a few healers also noted its use for 'yellow fever', which was later identified as hepatitis. Initial laboratory screening did not find any anticancer activity, but it did find that mamala extracts could protect cells against HIV-1. A particular isolated compound, prostratin (see box), was found to be responsible for the activity.

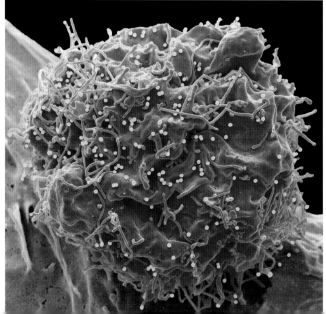

ART can bring down the levels of HIV in blood plasma to below the limits of detection but is effective only on actively replicating virus particles. If treatment is stopped for any reason, virus particles that are lying dormant (latent) in resting cells can become active and infect healthy cells. Prostratin can activate (induce) these dormant particles, which can then be targeted by the ART drugs. There has been much hope that this could result in a cure for HIV.

THE LONG ROAD

Cox, the AIDS Research Alliance and other organizations have agreed to share any profits from drugs that are developed from prostratin with the Government of Samoa, the Falealupo community and the healers who first told Cox about the use of mamala for hepatitis. Prostratin has undergone some clinical trials, but it takes many years for a potential drug candidate to become a licensed product, and the majority never do. Other sources of funding have been found to support the Falealupo village in the meantime, in order to provide infrastructure such as schools, protect the local rainforest and attract tourists who will themselves generate income through participating in activities such as a rainforest canopy walkway.

One of the limitations to commercialization of prostratin has been finding a cost-effective and reliable source. Efforts initially went towards cultivation of trees with the highest levels of the compound. In 2008, an economical method was found for synthesizing prostratin from phorbol, which can be obtained from croton oil (itself obtained from purging croton, *Croton*

ABOVE **Mamala (*Homalanthus nutans*), with a few fruit developing at the base of an old flower head.**

tiglium, also in the spurge family and currently already cultivated commercially). This method also enabled the prostratin structure to be modified to produce slightly different compounds. Some of these have been found to be a hundred times more effective than prostratin and are being investigated for their potential as HIV treatments.

Toxic relatives

Prostratin had already been discovered in the 1970s in the prostrate rice flower or Strathmore weed (*Pimelea prostrata*), a small shrub from New Zealand in the mezereum family (Thymelaeaceae). Some species of *Pimelea* have poisoned cattle and sheep, but this is primarily due to the presence of a daphnane-type diterpene ester called simplexin. Prostratin, on the other hand, is a diterpene ester of the phorbol type, many of which are irritant and promote tumour growth. Further research on mamala was therefore nearly halted, until Cox argued that traditional use suggested it was not acutely toxic and subsequent tests found that prostratin is not a tumour promoter.

ABOVE **Flowering prostrate rice flower (*Pimelea prostrata*) shrub.**

Malaria

Malaria is one of the biggest causes of disease worldwide, with almost half the global population at risk and more than 215 million cases and 445,000 deaths each year (some estimates put these figures much higher). Tropical and subtropical Africa bear the highest disease burden, and children are particularly vulnerable. The most effective treatments have originated from plant compounds, namely quinine from the bark of cinchona trees and, more recently, artemisinin from sweet wormwood.

ABOVE **Sweet wormwood (*Artemisia annua*) flowering plants have a long tradition of use to treat fever in China and now provide a valuable anti-malarial.**

NOBEL WINNER

PLANT:
Artemisia annua L.

COMMON NAME(S):
sweet wormwood, *qing hao*

FAMILY:
daisy (Asteraceae)

ACTIVE COMPOUND(S):

NATURALLY OCCURRING:
sesquiterpene lactones (artemisinin)

SEMI-SYNTHESIZED:
artemether; artesunate

MEDICINAL USES:

MAIN: anti-malarial

OTHER: fever

PARTS USED:
aerial flowering parts

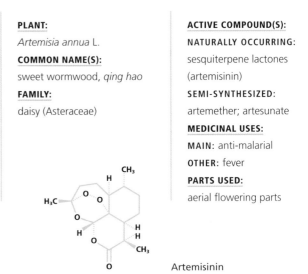

Artemisinin

ABOVE **The discovery of the sesquiterpene lactone artemisinin in sweet wormwood (*Artemisia annua*) and its anti-malarial properties resulted in the development of important new anti-malarial drugs.**

Sweet wormwood is an aromatic annual that can reach 1–2 m (3–6 ft) in height. It is native to North Africa, eastern Europe and much of Asia, and is widely naturalized elsewhere. Although it has been used in traditional Chinese medicine for several thousand years to treat fever and other conditions, it was not until the early 1970s that its potential as a treatment for malaria was investigated in the laboratory. At that time, the Chinese government was funding extensive studies into likely candidates from traditional Chinese medicine due to the growing resistance of the malaria parasite, a single-celled eukaryote, to drugs that were available at the time, many of which were based on quinine. The active

Fever tree

For 300 years from the mid-1600s, the powdered bark of cinchona trees (*Cinchona* spp.) from South America was an important remedy for fevers. Three species in particular were used medicinally, *C. pubescens*, *C. calisaya* (syn. *C. ledgeriana*) and *C. officinalis*, the first two of which became the main commercial sources. Jesuit missionaries in Peru noticed that the indigenous Quechua people used cinchona bark to relieve fever. It was soon being exported to Europe, where it was found to be particularly effective for treating fever associated with malaria, known then as the ague. The main active compound, quinine, was identified in 1820, making it one of the first plant alkaloids to be isolated.

Prolonged use of quinine can have serious side effects, so when the Second World War disrupted the supply of cinchona bark from plantations in Java, there was an added impetus to develop safer synthetic alternatives. Chloroquine was developed in the 1940s and became the most widely used preventative medicine. Unfortunately, the malaria parasites have become increasingly resistant to this and similar compounds, making the discovery of artemisinin from sweet wormwood particularly important.

LEFT Yellow cinchona (*Cinchona calisaya*), a small tree growing to 8 m (26 ft) tall, is native to Bolivia and Peru. Its bark was found to be one of the best sources of quinine.

compound in sweet wormwood was identified by the Chinese chemist Tu Youyou (b. 1930) as artemisinin (*qing hao su*), a sesquiterpene lactone. She went on to share the 2015 Nobel Prize in Physiology or Medicine for this discovery.

IRON-CLAD CURE

Since the discovery of artemisinin, clinical trials have confirmed its effectiveness in 90 per cent of malaria cases, including the most severe forms. The chemical structure of artemisinin includes a peroxide that reacts with iron in haem, the red pigment in blood cells, to generate highly reactive free radicals. When someone is bitten by a malaria-carrying mosquito, their red blood cells become host to the replicating parasite. If artemisinin enters the blood cell, the toxic free radicals are released, killing the malaria. In addition, semi-synthesized drugs based on artemisinin, such as artemether, have been developed and introduced, or are under investigation, as new forms of treatment in the arms race against malaria. These have largely replaced artemisinin as they have improved pharmaceutical properties. Unfortunately, however, the parasites are becoming resistant to artemisinin and its derivatives, so these are usually prescribed in combination with a quinine-derived anti-malarial to increase the effectiveness of the treatment.

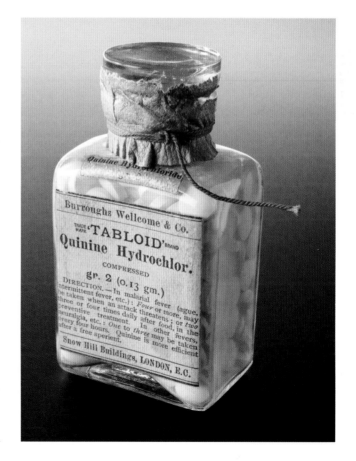

ABOVE Bottle of quinine hydrochloride tablets manufactured after 1895 by Burroughs Wellcome and Co., London, England.

CHAPTER 7

EASING MOVEMENT

There are many reasons why our movement may be affected. For example, it can be restricted by pain in our joints, and even certain disorders of the nervous system can affect our ability to control how we move. This chapter explores those plants and their constituents that can influence our movement in various ways.

Opium poppy (*Papaver somniferum*)

NATURAL COMPOUNDS FOR MUSCLES AND JOINTS

In some cases, joint pain might affect our ability to move. In other circumstances, dysfunction of parts of our nervous system can cause our movement to change or be difficult to control. Our ability to move may have various causes, but plant compounds can target many of these.

ABOVE **Flowering camphor tree (*Cinnamomum camphora*; see page 145) branch. The wood is a source of the ketone camphor, which is used externally for pain relief. Taking large doses of camphor internally or through the skin is now known to be harmful, and use of the compound is usually limited.**

A DOSE TO MAKE PAIN HISTORY

Some of the key pharmaceutical drugs for pain relief are derived from plants and have stood the test of time for their medicinal importance. These include painkilling analgesics from the opium poppy (*Papaver somniferum*; see page 140) in the poppy family (Papaveraceae), and aspirin, derived from chemicals in bark from willow (*Salix alba*) in the willow family (Salicaceae) and in meadowsweet (*Filipendula ulmaria*; see page 142) in the rose family (Rosaceae). Colchicine, in contrast, is a potent alkaloid from the autumn crocus (*Colchicum autumnale*; see page 148) in the family of the same name (Colchicaceae), and dampens the inflammatory response to relieve symptoms.

FOOD AND HERBS

Some plants that are familiar as foods and used for their delicious flavours also provide us with chemicals that are useful for pain relief. These include capsaicin from the chilli pepper (*Capsicum annuum*; see page 144) in the potato family (Solanaceae) and menthol from peppermint (*Mentha × piperita*; see page 145) in the mint family (Lamiaceae). These chemicals act in different ways to combat pain when applied as topical formulations to the skin.

Other plants traditionally used to ease pain are available as herbal medicines that contain mixtures of many different chemicals. One example is devil's claw (*Harpagophytum procumbens*; see page 146) in the sesame family (Pedaliaceae), whose tubers have been tested in humans for their potential usefulness in relieving pain caused by conditions such as osteoarthritis.

FROM AMAZON TO OPERATION

During certain surgical operations, it is important that the muscles are relaxed so that the patient does not move and disturb the procedure. Muscle-relaxant drugs are used by anaesthetists for this purpose. The origin of some of these drugs lies in plant-based arrow poisons, the 'curares' used traditionally in the Amazon

region to subdue victims and prey. The curare vine (*Chondrodendron tomentosum*; see page 152) in the moonseed family (Menispermaceae) is a source of the muscle-relaxing drug tubocurarine, which has inspired the design of other muscle-relaxant pharmaceuticals.

ABOVE **Chilli pepper (*Capsicum annuum*) fruit.**

MIND OVER MATTER

In some disorders, disruptions in neurotransmitters in certain parts of the brain can affect movement. For example, deficits in the neurotransmitter dopamine occur in Parkinson's disease and can influence the ability to move, or cause tremors. Some medicines restore the function of dopaminergic neurons in the brain; they include levodopa, which occurs in members of the legume family (Fabaceae), and apomorphine, which is derived from chemicals in the opium poppy (see page 155). Other plant chemicals restore the balance of nerve activity in the brain to alleviate Parkinson's symptoms, such as hyoscyamine from certain members of the potato family, including the thorn apple (*Datura stramonium*; see page 156).

Although this book focuses on plants, we cannot ignore the fact that fungi have also provided us with many useful drugs. Here, we give one example of how a fungus, ergot (*Claviceps purpurea*; see page 155) in the family Clavicipitaceae, has been instrumental in the development of drugs used for Parkinson's disease.

ABOVE **Flowering thorn apple (*Datura stramonium*) plant.**

Painkillers from plants

Some of the most widely used analgesic drugs today were originally derived from plant chemicals, including morphine and codeine from the opium poppy, and aspirin, derived from chemicals found in willow bark and other plants such as meadowsweet (see page 142). Some plants that are familiar to us as food provide different types of chemicals that are used to alleviate pain, including capsaicin from the chilli pepper (see page 144).

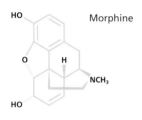

Morphine

LEFT **Morphine from the opium poppy (*Papaver somniferum*) was among the first alkaloids isolated from plants in the early nineteenth century.**

ABOVE **Opium poppy (*Papaver somniferum*) seed capsules are cut while they are still green so that the latex forms on the outside.**

POWERFUL POPPIES

PLANT:	MEDICINAL USES:
Papaver somniferum L.	MAIN: analgesic
COMMON NAME(S):	OTHER: sedative, to induce
opium poppy	euphoria
FAMILY:	PARTS USED:
poppy (Papaveraceae)	latex
ACTIVE COMPOUND(S):	
NATURALLY OCCURRING:	
morphinan alkaloids	
(morphine and codeine)	

When the unripe fruit capsules of the opium poppy are cut, milky-white droplets ooze out, turning brown and semi-solid on exposure to the air. This is opium, and its global trade and use have had an enormous impact on society throughout history, triggering wars and causing drug addiction, but also providing potent analgesics and a host of other useful medicines.

The word opium is derived from the Greek *opos*, meaning 'juice'. Its use dates back to ancient times – the Greek physician Hippocrates (*c.* 460–*c.* 370 BCE) was one of the first to recognize the medical applications of the plant. Historically, opium was used in rituals and was widely known for its sedative, euphoric and analgesic properties. In Victorian Britain, a tincture preparation of opium, called laudanum, was an analgesic and cough suppressant, although it could cause addiction.

In the early 1800s, morphine was isolated from the opium poppy. While it is an effective analgesic, it also causes euphoria and is narcotic. More than 20 years later, the related alkaloid codeine was isolated from opium; it has similar but less potent actions compared to morphine. Today, codeine is a widely used analgesic, available over the counter in tablet and other forms, while morphine is a potent analgesic prescribed for chronic pain. Both morphine and codeine interact with

ABOVE **Opium poppy (*Papaver somniferum*) flowers have four delicate petals, which may be white, pink or purple. In some countries the plants can be grown as ornamentals but commercial cultivation and opium harvesting can be carried out only with a licence due to the potential for the production of illegal drugs.**

HEROES OR FOES?

Opium contains another alkaloid, thebaine, first isolated in 1835. Although it is not a useful medicine itself, thebaine is chemically converted to codeine and other useful drugs to meet the demands of the pharmaceutical industry. Codeine, morphine and thebaine have been used to develop a wide range of other drugs, including the analgesics dihydrocodeine and buprenorphine, the cough suppressants pholcodine and dextromethorphan, and drugs such as methadone, used as substitution therapy in opioid dependence, including heroin addiction. Thebaine is also used to semi-synthesize naloxone, a drug that blocks opioid receptors and is useful in cases of opioid overdose to reverse the toxic effects.

There is, however, a darker side to opium: its use for the illicit manufacture of drugs such as heroin (a diacetyl derivative of morphine). At the beginning of the twentieth century, heroin was believed to be a safer drug than morphine and was prescribed to relieve pain and coughs. It was even named heroin because it was considered a 'heroic' drug. However, it soon became clear that heroin was highly addictive and so its use was restricted. Today, the manufacture, supply and possession of heroin and other opioid drugs is controlled by legislation.

opioid receptors in the nervous system. By mimicking the action of naturally occurring opioid peptides, they produce prolonged activation of these receptors to block pain, cause sedation and depress respiration. In the gut, this action can cause constipation. Because of these effects, morphine and codeine are sometimes used to alleviate diarrhoea (see page 95) and coughs.

Opium wars

In the Middle Ages, Arab and Turkish traders took opium to India and China, where it was used for medicinal applications. By the nineteenth century, smoking opium had become a popular practice in China and the Far East, and as demand for the drug increased, so did widespread addiction. The Chinese government attempted to combat the negative impact of opium on society, and in 1839 destroyed 20,000 opium chests exported to China by British merchants. Conflict erupted as a result, leading to the First Opium War. Peace was negotiated in 1842, but in the mid-1850s the Second Opium War began between China and France allied with the United Kingdom. This eventually resulted in the Treaties of Tianjin (1858), which legalized opium imports to China and expanded the number of ports open to foreign trade.

LEFT **Anglo-French troops invading Beijing, China, through the Tchao-yant Gate during the Second Opium War. Engraving by Jules Worms (1832–1934), published in *L'Illustration, Journal Universel*, Paris, 1860.**

A RIVERSIDE REVELATION

PLANT:

Salix alba L.

COMMON NAME(S):

willow, white willow

FAMILY:

willow (Salicaceae)

ACTIVE COMPOUND(S):

NATURALLY OCCURRING:

salicylates (salicin, salicylic acid)

SYNTHESIZED FROM NATURAL LEAD: aspirin

MEDICINAL USES:

MAIN: analgesic, anti-inflammatory, antipyretic

OTHER: skin disorders

PARTS USED:

bark

Salicylic acid

LEFT **Willow (*Salix* spp.) bark and meadowsweet (*Filipendula ulmaria*) herb contain salicylate chemicals, such as salicylic acid, which were the basis for the design of the analgesic and anti-inflammatory drug aspirin.**

BELOW **Willow (*Salix alba*) grows near streams and rivers, and is native to Europe, parts of North Africa and through Asia to northern China. These trees have deeply fissured bark (inset) and reach 25–30 m (80–100 ft) in height.**

Today, aspirin is a popular over-the-counter analgesic and anti-inflammatory medicine, and is also used as a preventative strategy in heart disease (see page 48), but the story of its discovery began more than 260 years ago. In 1757, Edward Stone (1702–1768), an English clergyman familiar with the doctrine of signatures, which states that natural remedies have an association with the conditions they are used to target, noticed that the 'ague' (the name at that time for malarial fever) was associated with marshy environments and their 'bad air' (*mal'aria* in Italian). This was before the discovery that mosquitoes transmit malaria (see page 134). It was during a riverside walk, therefore, that Stone searched for a remedy to cure his fever, leading him to select willow bark as one possible solution.

Stone found that regular doses of the powdered bark did indeed reduce his fever, so he gave this preparation to others with similar symptoms. Although they were not cured, their fevers were also alleviated. Stone published these results, which not only inspired herbalists to turn to willow bark as a fever remedy, but also stimulated scientific research to discover the active chemicals. In 1828, salicin, a phenolic glycoside, was isolated from willow bark. Two years later, the same compound was discovered in meadowsweet,

which also contains another salicylate, salicylic acid. When ingested, salicin is actually converted to salicylic acid, which is the active compound.

THE TEST OF TIME

In the nineteenth century, some doctors – including Thomas Maclagan (1838–1903) in Scotland, who also pioneered the use of clinical thermometers – prescribed salicin for rheumatic fever. However, there were concerns about the adverse effects of salicylates such as salicin, including gastric irritation and bleeding. Attempts were made to synthesize salicylate derivatives to retain the analgesic, anti-inflammatory and antipyretic (antifever) properties, while reducing gastric side effects. In 1897, treatment for pain relief was revolutionized when acetylsalicylic acid was synthesized on an industrial scale and marketed under the name aspirin – based on *Spiraea*, a genus in which meadowsweet has been placed previously.

Aspirin's mode of action was not discovered for another 80 years or so, when it was found to inhibit the enzyme cyclo-oxygenase, in turn resulting in the inhibition of inflammatory substances called prostaglandins. Pharmacologist John Vane (1927–2004) at London's Royal College of Surgeons shared the Nobel Prize in Physiology or Medicine in 1982 for his role in this important discovery. It could be said that discoveries arising from the willow and meadowsweet plants have truly made pain history.

ABOVE Meadowsweet (*Filipendula ulmaria*) is found in wet meadows, marshes and riverbanks throughout most of Europe, temperate Asia and parts of North America.

Willow the wisp

Ancient Egyptian texts record the use of willow bark for various conditions, while the ancient Greeks and Romans used it for pain and fevers. Hippocrates recommended a brew of willow leaves to ease pain during childbirth. The English herbalist Nicholas Culpeper (1616–1654) noted the use of willow 'To stay thin', and claimed it was 'Very good for redness and dimness of sight' and as a remedy 'To staunch bleeding of wounds, and at mouth and nose'.

The flowers of meadowsweet were used traditionally to treat inflammatory conditions such as rheumatism in the joints. The plant was a traditional remedy in the Scottish Highlands for fevers and headaches, and in Somerset for coughs when combined with parsley (*Petroselinum crispum*) in the carrot family (Apiaceae). As with remedies derived from willow, preparations using meadowsweet leaves were reputed to strengthen the eyes and prevent itching.

Many other plants also produce salicylates, so parts of these have been used for pain relief. They include leaves from wintergreen (*Gaultheria procumbens*) in the heather family (Ericaceae) and bark from sweet birch (*Betula lenta*) in the birch family (Betulaceae), both of which contain methyl salicylate. This form of salicylate is soluble in lipids (oils), so is formulated in liniments and rubs, and applied externally to the skin to relieve aches and pains, and occasionally chilblains.

ABOVE Wintergreen (*Gaultheria procumbens*) is a small evergreen shrub, native to eastern North America.

FEELING THE HEAT

PLANT:
Capsicum annuum L.
COMMON NAME(S):
chilli pepper, pepper
FAMILY:
potato (Solanaceae)

ACTIVE COMPOUND(S):
NATURALLY OCCURRING:
capsaicinoids (capsaicin)
MEDICINAL USES:
MAIN: pain relief (external)
OTHER: carminative
PARTS USED:
fruit

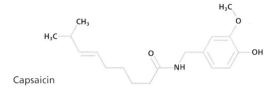

Capsaicin

ABOVE **Certain chilli peppers (*Capsicum annuum*) contain the pungent constituent capsaicin, which is applied to the skin as a rubefacient to alleviate joint pains.**

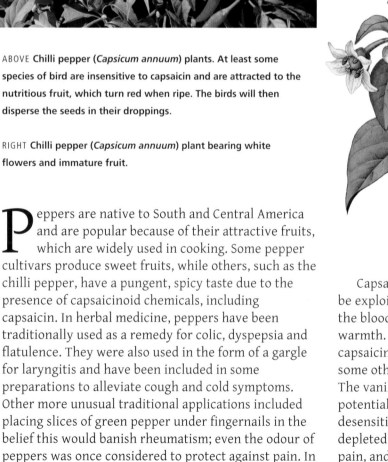

ABOVE **Chilli pepper (*Capsicum annuum*) plants. At least some species of bird are insensitive to capsaicin and are attracted to the nutritious fruit, which turn red when ripe. The birds will then disperse the seeds in their droppings.**

RIGHT **Chilli pepper (*Capsicum annuum*) plant bearing white flowers and immature fruit.**

Peppers are native to South and Central America and are popular because of their attractive fruits, which are widely used in cooking. Some pepper cultivars produce sweet fruits, while others, such as the chilli pepper, have a pungent, spicy taste due to the presence of capsaicinoid chemicals, including capsaicin. In herbal medicine, peppers have been traditionally used as a remedy for colic, dyspepsia and flatulence. They were also used in the form of a gargle for laryngitis and have been included in some preparations to alleviate cough and cold symptoms. Other more unusual traditional applications included placing slices of green pepper under fingernails in the belief this would banish rheumatism; even the odour of peppers was once considered to protect against pain. In Trinidad, the leaves have been used as a traditional remedy for asthma.

Capsaicin can irritate the skin, but this action can be exploited for therapeutic purposes as it increases the blood circulation in the skin and creates a feeling of warmth. This burning-like sensation occurs because capsaicin activates vanilloid receptors in nerves and in some other tissues, including skin cells (keratinocytes). The vanilloid receptors, such as transient receptor potential vanilloid type 1 (TRPV1), are believed to be desensitized by capsaicin. Sensory nerve fibres are also depleted of a peptide (substance P) that is involved in pain, and this contributes to the analgesic effect. These actions explain why capsaicin is included in some liniments and other preparations that are applied to the skin to relieve joint pains, such as in osteoarthritis, and to alleviate some types of nerve pain, such as post-herpetic neuralgia following shingles.

Today, the oleoresin from *Capsicum* peppers has been developed for a very different purpose that also relies on the irritant and burning sensation of the capsaicinoid constituents: in 'pepper sprays' for law enforcement or self-defence.

FROM TREE TO TREATMENT

Some other plant chemicals also produce skin irritation, so when applied they act as counterirritants, helping to counteract pain signals; they are known as rubifacients. One example is camphor, a monoterpene ketone that occurs in plants such as rosemary (*Salvia rosmarinus*, syn. *Rosmarinus officinalis*) in the mint family (Lamiaceae) and contributes to the characteristic fragrance of the leaves. Camphor is usually obtained from wood of the camphor tree (*Cinnamomum camphora*; see page 138) in the laurel family (Lauraceae), which has a native range extending from north Vietnam to west-central and south Japan. Camphor is used as a rubifacient and mild analgesic in liniments and other preparations that are applied externally to the skin to relieve some types of pain such as neuralgia. It is also sometimes included in certain decongestant remedies in the form of an inhalation.

A cooling sensation

One of the main constituents of peppermint oil is the monoterpene menthol, which produces a sensation of cold and a mild analgesic effect when applied to the skin (see page 90). This effect is caused by the menthol acting on transient receptor potential melastatin type 8 (TRPM8) channels, temperature-sensitive receptors that are activated by low temperatures. Menthol is therefore applied to the skin in topical preparations to relieve itching (pruritus), such as that caused by urticaria. When diluted in topical preparations, menthol – either alone or as a component of peppermint oil – has also been of interest to alleviate some types of pain. For example, certain menthol preparations have been investigated for their usefulness in relieving headaches when applied to the forehead, and in post-herpetic neuralgia. However, more studies are needed to explore any potential benefits in these conditions.

LEFT **Rosemary (*Salvia rosmarinus*) is an aromatic shrub, native to the countries of Europe and North Africa bordering the Mediterranean, and is widely cultivated elsewhere. It has a long history of use in medicine, to flavour food and in perfumes.**

The root of the pain

Some plants traditionally used for pain relief, such as devil's claw, are still popular as herbal medicines, although they have not been investigated as extensively as plant-derived drugs. Plants in our diet have also been of interest for any role they may play in protecting against certain conditions that lead to joint pain.

DEVIL'S CLAW

PLANT:

Harpagophytum procumbens (Burch.) DC. ex Meisn.

COMMON NAME(S):

devil's claw, grapple plant

FAMILY:

sesame (Pedaliaceae)

ACTIVE COMPOUND(S):

NATURALLY OCCURRING:

iridoid glycosides (harpagoside)

MEDICINAL USES:

MAIN: analgesic

OTHER: stomachic

PARTS USED:

root tubers

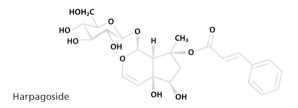

Harpagoside

ABOVE **Iridoid glycoside constituents, including harpagoside, may contribute to the suggested pain-relieving effects of devil's claw (*Harpagophytum procumbens*) root tuber extracts.**

ABOVE **Slices of devil's claw (*Harpagophytum procumbens*) root tuber, for medicinal use.**

Devil's claw gets its name from its spiny fruits, but it is the root tubers that have been of interest for rheumatic conditions. They contain several groups of chemicals, including iridoid glycosides, flavonoids, triterpenes and phenolic acids. Some laboratory studies into root extracts have indicated that they have anti-inflammatory and analgesic effects, although the compounds responsible for these actions require further investigation. Certain studies suggest that the iridoid glycoside harpagoside and its aglycone, harpagogenin, are anti-inflammatory. Scientific research has revealed that extracts containing harpagoside inhibit the production of some inflammatory mediators, indicating they may act in a similar way to aspirin, by inhibiting the enzyme cyclo-oxygenase.

Root tuber extracts have also been shown to inhibit the production of other inflammatory mediators called cytokines, such as interleukin-1, *beta* (IL-1β) and tumour necrosis factor-*alpha* (TNF-α). These cytokines not only contribute to inflammation, but they also increase the production of enzymes called matrix metalloproteinases (MMPs), which break down cartilage, suggesting devil's claw might also help protect this tissue in arthritic joints. When tested in clinical studies in humans with either osteoarthritis, lower back pain or other painful conditions, certain devil's claw preparations could reduce pain when compared to a placebo. However, not all studies in humans have had such promising results and more research is needed to confirm any benefits devil's claw may have in alleviating pain.

FOOD FOR FLEXIBILITY

Wild cabbage (*Brassica oleracea*) in the cabbage family (Brassicaceae) is native to the Mediterranean region and southwestern Europe. The species has been cultivated for at least 2,000 years and has given us various edible cultivars, including cabbage, kale, cauliflower, broccoli and Brussels sprouts. Some cultivars, such as broccoli, contain isothiocyanate chemicals, including sulforaphane, which can reduce the expression of MMPs and so may help protect the cartilage in joints. As sulforaphane also blocks the action of some inflammatory mediators, it is of interest for use in osteoarthritis. Other plants eaten as part of our diet – including garlic (*Allium sativum*) and other members of the onion genus (*Allium* spp.) in the amaryllis family (Amaryllidaceae), green tea (*Camellia sinensis*; see pages 62 and 66) in the tea family (Theaceae) and turmeric (*Curcuma longa*; see pages 106 and 216) in the ginger family (Zingiberaceae) – have also been investigated for their potential usefulness against the progression of osteoarthritis.

Protecting the protector

In the Kalahari Desert, devil's claw has been used for centuries as a remedy for conditions ranging from digestive and blood disorders to infections. The plant was first brought to Europe in the 1950s and has since gained a reputation for alleviating rheumatic complaints, but it has also been used as a bitter tonic. In southern parts of Africa, local people harvest the roots and sell them for export to support their livelihoods. Most of the world's supply of devil's claw comes from Namibia, with lesser amounts from South Africa and Botswana. Devil's claw is now a protected species in these three countries and its trade is regulated under the Convention on International Trade in Endangered Species of Wild Fauna and Flora, an agreement between governments aimed at ensuring the international trade in specimens of wild plants does not threaten their survival.

ABOVE **Devil's claw (*Harpagophytum procumbens*) fruit (inset) are dispersed by hooking onto the fur of animals.**

Disease of kings

The condition known as gout is associated with raised levels of uric acid in the blood, which can be deposited in the joints and lead to inflammation. One of the drugs used for attacks of gout is colchicine, an alkaloid that occurs in the autumn crocus from Europe and in the flame or glory lily (*Gloriosa superba*), another plant from the same family but found in southern Africa and tropical Asia.

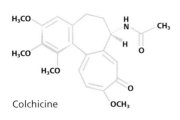

LEFT The alkaloid colchicine is restricted to members of the autumn crocus family (Colchicaceae). It has been developed as a pharmaceutical to alleviate symptoms of gout.

Colchicine

ABOVE Autumn crocus (*Colchicum autumnale*) flowers usually appear in autumn, after the leaves have died back, giving rise to the plant's alternative common name of naked ladies.

THE AUTUMN CROCUS

PLANT:
Colchicum autumnale L.
COMMON NAME(S):
autumn crocus, meadow saffron, naked ladies
FAMILY:
autumn crocus or naked-ladies (Colchicaceae)

ACTIVE COMPOUND(S):
NATURALLY OCCURRING:
phenethylisoquinoline alkaloid (colchicine)
MEDICINAL USES:
MAIN: gout
OTHER: expectorant
PARTS USED:
seeds and corms

The autumn crocus is native to parts of Europe, including Britain, France and Germany, although it has been introduced to other regions, ranging from Denmark to the South Island of New Zealand. The genus name *Colchicum* is derived from Colchis, an ancient region on the eastern edge of the Black Sea, where the plant grows. The plant has a long-standing reputation for being poisonous, although there are some records of its traditional use for medicinal purposes (see box).

In 1820, the alkaloid colchicine was first isolated from the autumn crocus, and although it has since been of interest for use in leukaemia, it is currently used for gout. Colchicine is not an analgesic and has only weak anti-inflammatory activity, so it targets gout symptoms by other mechanisms. The alkaloid binds to tubulin in some white blood cells (leucocytes) called neutrophils, and inhibits their activation and migration to areas where uric acid has deposited, such as joints. This action reduces the inflammatory response and so alleviates gout symptoms. Colchicine is not only used for gout, but also for another condition called familial Mediterranean fever, which involves inflammation of the joints and some other parts of the body. Commercial sources of colchicine for the development of medicines by the pharmaceutical industry include seeds from autumn crocus plants cultivated in parts of Europe, such as the Netherlands, and the flame lily, grown in India.

A DOSE OF SHERRY

Autumn crocus corms were included as a medicine in the London Pharmacopoeias of the seventeenth century and, although the use of corms later declined, the seeds were introduced as a medicinal substance in nineteenth-century London Pharmacopoeias. In the very first edition of the British Pharmacopoeia, published in 1864, both the corm and seed of the autumn crocus were included. The book described how to prepare a tincture made from the seeds, and how to produce a medicinal wine by macerating dried and sliced autumn crocus corms in sherry.

BELOW **Glory lily (*Gloriosa superba*) is a tropical climbing plant from parts of Asia and Africa. The scientific name of the genus means 'full of glory' and is inspired by the striking flowers.**

Poisonous and powerful potions

Pedanius Dioscorides (*c.* 40–90 CE), a Greek physician, was familiar with the poisonous properties of *Colchicum* and it was still much avoided in medieval times for this reason, although ancient Arabic texts suggested it had uses for gout. In France, autumn crocus flowers were crushed and placed on the heads of children in the belief that doing so would destroy vermin. An Anglo-Saxon remedy for facial pimples recommended combining autumn crocus corms with oil and using the preparation as a wash.

Nicholas Culpeper described the poisonous actions of ingesting a single 'grain' (seed) of the autumn crocus, saying that doing so caused 'Flushes of heat and shiverings, colicky pains and irritation in the loins and urinary passages.' However, the English herbalist also wrote that, when prepared properly, autumn crocus was a powerful medicine for 'Dropsies and tertian agues', that when taken as a syrup preparation it 'Gently bites the tongue and is excellent for cleansing it from mucus', and that it had potent expectorant properties.

CENTRAL AND SOUTH AMERICA

The combined region of Central and South America stretches from Mexico and the Caribbean islands in the north, to Argentina and Chile in the south. It includes the rainforest of the Amazon Basin, the high mountains of the Andes and the deserts of Mexico, along with many other habitats. An estimated 24,000 plant species occur in Mexico and 46,000 species in Brazil, although these figures will likely increase as work continues to identify the plants of this rich and diverse region.

Civilizations and influences

Prior to the arrival of Italian explorer Christopher Columbus (1451–1506) in the Caribbean in 1492 and the subsequent conquest by the Spanish of Central and South America, there were three main civilizations in the region: the Aztecs in Mexico; the Maya from southern Mexico to Guatemala, Belize, Honduras and Costa Rica; and the Incas from Peru to Chile and Argentina. Unfortunately, little evidence exists of the medico-religious systems of these civilizations or of the smaller groups of indigenous people who lived in other parts of the region.

Early European settlers – predominantly from Spain, Portugal and France – promoted Christianity but allowed local populations to continue practising their medicine and

were interested in learning from them, taking some plants back with them to Europe (see box). In turn, European medicinal plants have been incorporated into herbal medicinal practice in the region. Later influences came from African slaves, who were transported across the Atlantic to work on plantations and who brought with them their own plants and rituals. Some medicinal plants from India are now also used.

Potato family favourites

A number of members of the potato family that are valued around the world originated in Central and South America. The potato (*Solanum tuberosum*) itself is probably native to southern Peru, from where ceramic vessels based on the potato tuber and dating to the sixth century CE were associated with the worship of the deity Axomama, mother of the potato. Potatoes were being cultivated in temperate regions from Chile to Colombia by the sixteenth

BELOW **Potato (*Solanum tuberosum*) is a South American plant but is now cultivated as a staple root crop for many people around the world.**

century, and another plant native to Peru, the tomato (*S. lycopersicum*; see page 216), was being cultivated in Mexico. Early illustrations of tomatoes dating from shortly after the species' introduction to Europe show that the fruit already had a range of colours (white or yellow to red) and shapes (small and round to large and ribbed).

Members of the genus *Capsicum*, which includes the chilli pepper (*Capsicum annuum*; see page 144), are native from Mexico to northern Argentina. In 1493, five months after he first landed on the Caribbean island of Hispaniola, Columbus wrote in his diary about chilli, which he called *axi*, saying that it was like pepper and 'everybody does not eat without it, it being very healthy'.

Somewhat less healthy is tobacco (*Nicotiana tabacum*; see page 68) from Bolivia, which, together with Aztec tobacco (*N. rustica*) from Peru, was widely used in many areas of Central and South America by the sixteenth century. Leaves of the plants were smoked, chewed and taken as snuff as a medicine and during rituals. Both species were introduced to Europe, as sweet and coarse tobacco, respectively, where they were used as a panacea.

RIGHT **Flowering Aztec tobacco (*Nicotiana rustica*) plants. The leaves are a source of the addictive alkaloid nicotine.**

CODEX DE LA CRUZ-BADIANO

The oldest surviving written record of herbal medicine in the region is the *Codex de la Cruz-Badiano*, also known by various other names, including *Libellus de Medicinalibus Indorum Herbis* (*Little Book of the Medicinal Herbs of the Indians*). It was written in 1552 by two Aztec scholars who were educated in Santa Cruz, Mexico, and about whom very little is otherwise known. The main author, Martín (Martinus) de la Cruz, wrote it in the Aztec language Nahuatl, and it was translated into Latin by Juan Badiano (Juannes Badianus). The Latin translation survives and, after spending most of the intervening time in Europe, was returned to Mexico in 1992 by Pope John Paul II (1920–2005). The codex covers more than 200 plants and shows that pre-Conquest Aztec botany

and medicine were both sophisticated systems. It illustrates and describes the uses of plants that influenced European medicine, food and gardens, such as *tlapalcacauatl* or chocolate (*Theobroma cacao*; see page 31) in the mallow family (Malvaceae), *tlilxochitl* or vanilla (*Vanilla planifolia*) in the orchid family (Orchidaceae), and *cohuanenepilli*, a red-flowered dahlia (probably *Dahlia coccinea*), in the daisy family (Asteraceae).

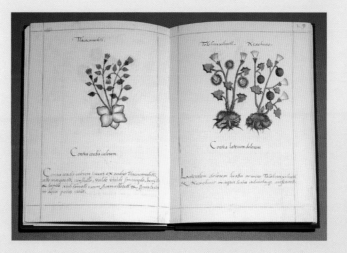

LEFT ***Codex de la Cruz-Badiano* facsimile. The page on the right shows two species of *Datura* – *tolohua xihuitl* (*D. innoxia*) and *nexehuac* (*D. ceratocaula*) – with the advice that they should be applied for a pain in the side. The illustrations show the spiky fruit of the former and smooth fruit of the latter.**

Manipulating the muscles

For some types of surgical operations, the patient's muscles are relaxed using drugs to prevent them from moving during the procedure. Some muscle-relaxant drugs are derived from plants, including tubocurarine from the curare vine. Other plants and their constituents are used to relieve muscle spasticity in conditions such as multiple sclerosis, and in recent years medicinal products derived from cannabis (*Cannabis sativa*) in the cannabis family (Cannabaceae) have been developed for this use.

FROZEN IN TIME

PLANT:

Chondrodendron tomentosum Ruiz & Pav.

COMMON NAME(S):

curare vine, pareira brava

FAMILY:

moonseed (Menispermaceae)

ACTIVE COMPOUND(S):

NATURALLY OCCURRING:

bisbenzylisoquinoline alkaloid (tubocurarine)

SYNTHESIZED FROM NATURAL

LEAD: atracurium

MEDICINAL USES:

MAIN: skeletal muscle relaxant

OTHER: arrow poison

PARTS USED:

stems

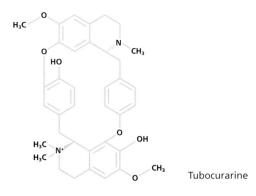

Tubocurarine

ABOVE **The curare vine (*Chondrodendron tomentosum*) contains the alkaloid tubocurarine, which relaxes muscles. It has been used by anaesthetists during surgery and has inspired the design of new muscle-relaxant drugs that are used pharmaceutically today for the same purpose.**

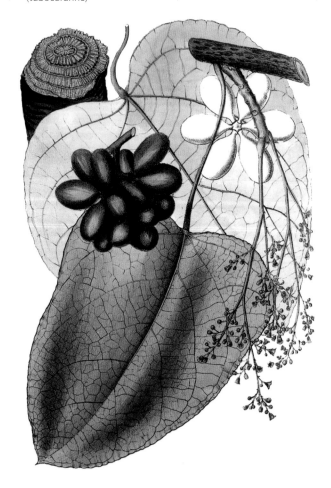

In the Amazon region of South America, different plants have been used to make arrow poisons, collectively known as 'curare' (see box). One such plant is the curare vine. Its stems contain the alkaloid tubocurarine, which blocks the action of the neurotransmitter acetylcholine at the junction between nerve and skeletal muscle cells. This interferes with the communication between these cells, causing the skeletal muscles to relax. This action explains why arrows tipped in preparations made from the curare vine could have been effective in paralyzing the muscles of victims and prey, and also led to the use of tubocurarine as a drug administered by anaesthetists to relax muscles during surgery. Tubocurarine has also sometimes been used to control muscle spasms caused by tetanus, which is a bacterial infection of the nervous system.

LEFT **Curare vine (*Chondrodendron tomentosum*) is a woody stemmed climbing plant from northwest South America. It was traditionally used as an arrow poison.**

Tubocurarine can, however, have unwanted effects, such as lowering blood pressure. It can also cause the release of histamine, which has been linked with hypersensitivity (allergic-type) reactions. Tubocurarine has therefore largely been replaced in surgical procedures with other muscle-relaxing drugs that have improved pharmaceutical properties, although its chemical structure inspired the design of some of these, including atracurium.

MEDICINAL CANNABIS

Multiple sclerosis affects the nerves in the brain and spinal cord, resulting in symptoms that can include spasticity of the muscles, which is an increase in muscle tone. Cannabis (also known as marijuana and many other names) contains cannabinoid constituents, including the psychoactive *delta*-9-tetrahydrocannabinol (THC or dronabinol; see also page 83) and cannabidiol, which act on cannabinoid receptors in the nervous system to modulate the actions of neurotransmitters. A mixture of these two cannabinoids has been developed as a medicine to relieve symptoms of muscle spasticity in multiple sclerosis, while cannabidiol is being studied for its potential usefulness in some types of seizures.

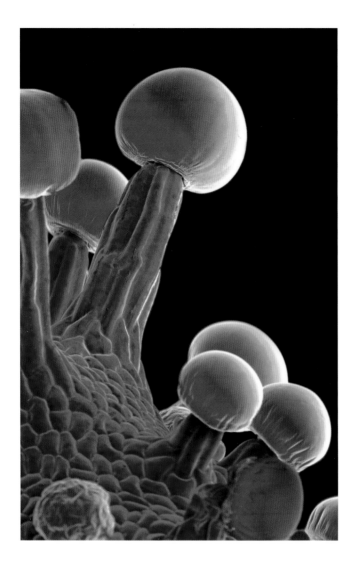

LEFT Colour-enhanced scanning electron micrograph of the surface of a cannabis (*Cannabis sativa*) leaf. It shows the glandular cells (trichomes) in which cannabinoids accumulate.

Arrow poisons

South American arrow poisons (curares) have traditionally been prepared from plants, including some that are poisonous, and aim to subdue hunting prey. A number of arrow poisons (particularly those used in Brazil and Peru) are from plant species in the moonseed family, including the curare vine, a source of tubocurarine (see main text), and species of *Cissampelos*. Scientific studies have revealed that velvet leaf or midwives' herb (*C. pareira*) contains bisbenzylisoquinoline alkaloids such as hayatin, which has a similar muscle-relaxing action to tubocurarine. This might explain the traditional use of *Cissampelos* species in arrow poisons, and perhaps also in traditional remedies in Central America to calm and relax victims of snakebites.

Elsewhere in South America, including Colombia and Venezuela, curares have been traditionally prepared from other plants, especially species of *Strychnos* in the Indian pink family (Loganiaceae). Devildoer (*S. toxifera*) was an ingredient of calabash curares. It contains indole alkaloids that can cause paralysis, including toxiferine, and has been developed into the muscle-relaxing drug alcuronium. *Strychnos* species such as nux vomica (*S. nux-vomica*) from India and Malaysia are poisonous, principally due to the presence of strychnine. This indole alkaloid also affects the nervous system, causing tremors and convulsions. Strychnine is now mainly used as a rodenticide, but it was formerly used for some eye and urinary tract disorders, and to stimulate the nervous system.

Parkinson's disease

In Parkinson's disease and other movement disorders, current therapies aim to modulate the actions of neurotransmitters in the brain to help relieve symptoms. Some medicines developed for these conditions occur in plants such as the velvet bean (*Mucuna pruriens*) in the legume family, or have been developed from chemicals originally from plants such as the thorn apple in the potato family and ipecacuanha (*Carapichea ipecacuanha*) in the coffee family (Rubiaceae) (see page 156).

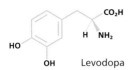

LEFT **Levodopa is used pharmaceutically to relieve some symptoms in Parkinson's disease.**

THE VELVET BEAN

PLANT:
Mucuna pruriens (L.) DC.

COMMON NAME(S):
velvet bean

FAMILY:
legume (Fabaceae)

ACTIVE COMPOUND(S):
NATURALLY OCCURRING:
amino acid (levodopa [L-DOPA])

MEDICINAL USES:
MAIN: Parkinson's disease
OTHER: other neurological disorders

PARTS USED:
seeds (beans)

In Parkinson's disease, certain types of neurons (especially dopaminergic neurons) degenerate in the brain and so the levels of the neurotransmitter dopamine are depleted. This occurs in parts of the brain that have an important role for movement, explaining why, in Parkinson's disease, movement disorders such as tremors occur. Current therapies aim to restore the communication between dopaminergic neurons to reduce these symptoms. One of these drugs is levodopa, which can cross the blood–brain barrier, a protective semi-permeable membrane around the brain that allows only certain substances to pass across it. Once levodopa reaches the brain, it is converted by an enzyme into dopamine, restoring levels and alleviating Parkinson's symptoms.

It is now known that seeds of some plants in the legume family, including the velvet bean and the broad or faba bean (*Vicia faba*), also contain levodopa. This could explain why the velvet bean has been used in traditional Ayurvedic medicine for movement disorders, including for symptoms of Parkinson's. In laboratory studies, extracts of velvet bean containing levodopa have been shown to improve the function of dopaminergic neurons. Powdered velvet beans have also been tested in people with Parkinson's disease in

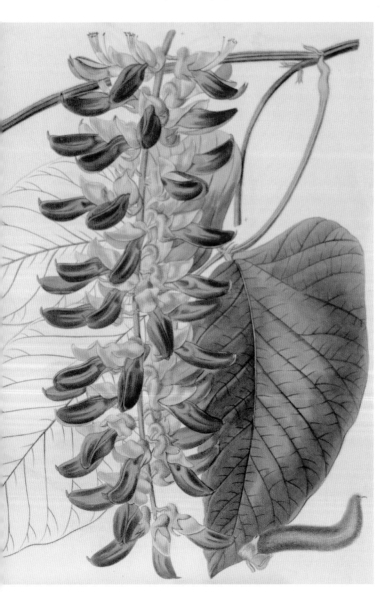

ABOVE **Velvet bean (*Mucuna pruriens*).**

Fungus fiend becomes friend

In northern Europe in the Middle Ages, cases of mass poisoning through eating bread made with flour from rye (*Secale cereale*), a member of the grass family (Poaceae), were common. The rye was contaminated with the ergot fungus, which can infect the ears of this and related cereal crops. It resulted in symptoms of ergotism, including convulsions, hallucinations, gangrene and death. The Order of Hospitallers of St Anthony in France became a place of pilgrimage for those afflicted with ergotism, so the condition also became known as St Anthony's fire.

The poisonous properties of the ergot fungus are due to the indole alkaloids it contains, including ergotamine and ergometrine. These alkaloids constrict blood vessels to reduce blood flow and can produce uterine contractions, so they have found medicinal uses for some types of headaches (ergotamine) and in childbirth (ergometrine). Modification of the chemical structures of these alkaloids has produced new drugs such as bromocriptine, which bind to dopamine receptors more selectively. Such ergot-derivative drugs can therefore mimic the action of dopamine in the brain and are used to alleviate symptoms in Parkinson's disease.

ABOVE Rye (*Secale cereale*) infected with the ergot fungus (*Claviceps purpurea*) in the family Clavicipitaceae.

some small clinical studies to assess any potential benefits, although the results were variable and therefore inconclusive.

A POPPY'S NEW PURPOSE

We are most familiar with the opium poppy as a source of the analgesics codeine and morphine (see page 140). However, back in 1869 it was first reported that chemical modification could convert morphine into apomorphine, and more recently this chemical has been synthesized from thebaine, another compound in the opium poppy. While apomorphine cannot bind effectively to opioid receptors and so is not useful as an analgesic, it can bind to dopamine receptors and so can mimic the action of the neurotransmitter dopamine. In the 1950s, the first reports emerged that apomorphine could alleviate Parkinson's, and today, it is one of the medicines prescribed to treat symptoms of this disease.

LEFT Velvet bean (*Mucuna pruriens*) is an attractive liana from tropical and subtropical Africa and Asia. The velvet hairy coating to the pods gives rise to its common name.

A POISONOUS APPLE

PLANT:

Datura stramonium L.

COMMON NAME(S):

thorn apple, jimsonweed,
Jamestown weed,
stramonium

FAMILY:

potato (Solanaceae)

ACTIVE COMPOUND(S):

NATURALLY OCCURRING:

tropane alkaloids
(hyoscyamine, hyoscine)

SYNTHESIZED FROM NATURAL

LEAD: benzatropine
[benztropine]

MEDICINAL USES:

MAIN: Parkinson's disease

OTHER: asthma

PARTS USED:

leaves

(–)-Hyoscyamine

ABOVE **Certain members of the potato family, including the thorn apple (*Datura stramonium*), contain tropane alkaloids, including the pharmacologically active (–)-hyoscyamine. This can be converted to the (+)-form, giving a mixture known as atropine, which is of interest for pharmaceutical applications.**

Thorn apple was used as a traditional remedy in North America and was known to have intoxicating and hallucinogenic properties (see box). Along with some other members of the potato family such as black henbane (*Hyoscyamus niger*; see page 86) and belladonna (*Atropa bella-donna*; see page 180), thorn apple contains tropane alkaloids such as hyoscyamine, which block the action of the neurotransmitter acetylcholine at cholinergic receptors on nerve cells.

In Parkinson's disease, it is thought that, as the activity of the neurotransmitter dopamine declines in the brain, the activity of other types of neurons (cholinergic neurons) become overactive in comparison. One strategy to alleviate symptoms in Parkinson's disease is to restore the balance between the less active dopaminergic neurons and the relatively more active cholinergic neurons. Drugs that block cholinergic receptors to dampen their activity, such as the anticholinergic hyoscyamine, have therefore been used for this purpose.

Hyoscyamine can occur as two different chemical forms, (–)-hyoscyamine and (+)-hyoscyamine, and the mixture of both is known as atropine. Newly developed drugs based on the chemical structure of atropine, including benzatropine, have also sometimes been used as anticholinergic drugs to manage Parkinson's symptoms.

IPECACUANHA

Ipecacuanha has a native range extending from southeast Nicaragua to Brazil and was introduced to Europe in the seventeenth century. The roots have long been prepared as an expectorant for coughs, as an emetic and for dysentery, although these uses have generally been discontinued due to safety concerns. In 1817, emetine, an isoquinoline-derived alkaloid, was

first isolated from ipecacuanha roots, and in the 1960s it was used as the lead compound in the development of the drug tetrabenazine for some types of nerve disorders. Tetrabenazine inhibits the uptake of some neurotransmitters in the brain, thereby depleting the stores available in nerve cells. In 2008, the United States Food and Drug Administration approved use of the drug for involuntary movements (chorea) in Huntington's disease, and it has been investigated for use in some other movement disorders that have a neurological component.

Historical hallucinogen

In the 1600s, with finance from the Virginia Company in London following a charter granted to investors by King James I of England and Ireland, English colonists established the settlement of Jamestown in Virginia, North America. Soon after, there were reports of some settlers being poisoned by the thorn apple, although the plant was later used for its medicinal properties. In 1676, British soldiers were sent to Jamestown to halt the rebellion of Nathanial Bacon (1647–1676). Not knowing that the thorn apple was poisonous, the soldiers consumed its leaves and experienced hallucinogenic effects, getting up to all sorts of antics as a result. The role of the thorn apple in Jamestown's history is the origin of two of the plant's other common names, Jamestown weed and jimsonweed.

EASING MOVEMENT

ABOVE **A form of thorn apple or jimsonweed (***Datura stramonium***,
syn.** ***D. tatula***) with lilac flowers and colouring to the stems, bearing
the typical spiny fruit of the species.**

CHAPTER 8

HEALTHY ON THE OUTSIDE

Our skin acts as a barrier to many harmful substances and has other important functions, including regulating our body temperature. While it plays a vital role in keeping us healthy, skin can be damaged by wounds and injuries, or affected by disease, infections and parasites. Skin is also crucial to our sense of touch, while another of our senses – sight – is mediated by our eyes.

Greater ammi (*Ammi majus*)

NATURAL PRODUCTS FOR THE SKIN AND EYES

Our skin may be affected by different diseases, ranging from wounds and inflammation to infections and infestations, and our eyes may also become afflicted by disease. The medicines used for skin and eye disorders vary considerably, depending on the cause. Plants produce extremely diverse chemicals, so it is not surprising that some have been developed as useful pharmaceuticals, or are of interest as traditional herbal remedies for skin and eye problems.

ABOVE **Leafy branch of the Peru balsam tree (*Myroxylon balsamum*), with racemes of flowers, developing pods (fruit) and a single mature pod. Peru balsam itself is obtained after fire-scorching the tree bark and is collected from both the bark, once removed, and the exposed trunk. Tapping the tree produces tolu balsam, which has a different chemical composition.**

HERBAL HELPERS FROM HISTORY

Throughout history, many plants have been reputed to aid skin conditions, such as wounds, burns, bites and stings. Of those that have been investigated scientifically, there does appear to be a rational basis for the use of some. For example, aloe vera (*Aloe vera*; see page 164) in the asphodel family (Asphodelaceae) and gotu kola (*Centella asiatica*; see page 166) in the carrot family (Apiaceae) have been used historically to soothe inflamed or wounded skin. Arnica (*Arnica montana*; see page 166), a member of the daisy family (Asteraceae), is reputed to relieve bruises and sprains. These and other plants are still used as traditional herbal remedies for skin complaints, and we now have some knowledge of their chemical constituents and how they may act on the skin to relieve symptoms.

PLANT-DERIVED PHARMACEUTICALS

Widely used pharmaceutical drugs in modern medicine include anti-inflammatory corticosteroids that are applied to certain skin conditions. The starting material to manufacture these drugs can be sourced from steroidal compounds in plants such as sisal (*Agave sisalana*; see page 162) in the asparagus family (Asparagaceae). Other plants contain chemicals that have inspired the design of new pharmaceuticals. For example, a chemical in Goa powder, obtained from the Brazilian araroba tree (*Vataireopsis araroba*, syn. *Andira araroba*) in the legume family (Fabaceae), and psoralens that occur in plants such as the greater ammi (*Ammi majus*) in the carrot family, resulted in the development of topical medicines for psoriasis (see page 168).

INFECTIONS AND INFESTATIONS

Our skin may be vulnerable to infections caused by bacteria, fungi or viruses, or parasites that live on our skin. Plants produce chemicals for their own survival, some of which protect them from damage caused by infections or deter predators such as insects. Therefore, many compounds with antimicrobial properties, or insect-repellent or even insecticidal actions, can be found in plants. This explains why some plants have been studied for their usefulness against human skin infections and parasites, including the essential oil (or volatile oil) from tea tree (*Melaleuca alternifolia*; see page 170) in the myrtle family (Myrtaceae). Other plant constituents can help destroy or remove infected skin cells, such as those occurring in the rhizomes of mayapple (*Podophyllum peltatum*) in the barberry family (Berberidaceae) and willow bark (*Salix alba*) in the willow family (Salicaceae) (see page 142). Benzyl benzoate, a chemical that occurs in Peru balsam – obtained from the Peru balsam tree (*Myoxylon balsamum*) in the legume family – has activity against scabies mites (*Sarcoptes scabiei*), so has been used as a treatment to eradicate these skin parasites (see page 176).

A SIGHT FOR SORE EYES

Particular plant alkaloids can influence the function of parts of our nervous system, so have been developed to control the muscles in the eyes that can relieve symptoms of some conditions. For example, alkaloids from Maranham jaborandi (*Pilocarpus microphyllus*) in the citrus family (Rutaceae) and the Calabar bean (*Physostigma venenosum*) in the legume family have been used in the form of drops for glaucoma (see page 178). Another alkaloid from belladonna (*Atropa bella-donna*) in the potato family (Solanaceae) was once applied to the eyes as a traditional cosmetic, but now has a medical use in ophthalmology (see page 180).

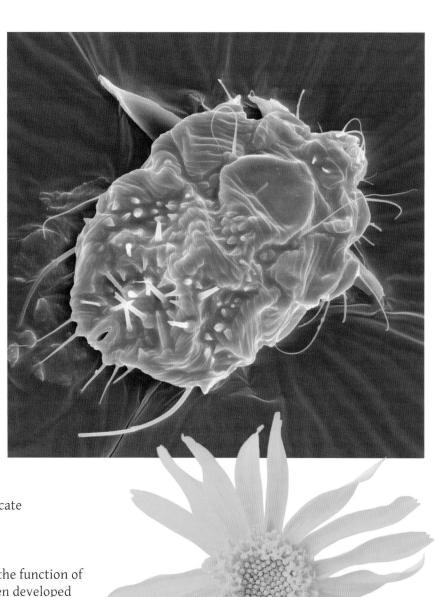

ABOVE **Scanning electron micrograph of a greatly magnified scabies mite (*Sarcoptes scabiei*). These mites burrow into skin and cause an intensely itchy rash.**

RIGHT **Extracts of arnica (*Arnica montana*) flower heads are used traditionally on the skin as an anti-inflammatory.**

Skin solutions

Our skin provides a barrier, protecting our internal organs, but as a consequence it may be damaged itself. Some plants have long been used to alleviate skin injuries and wounds, including aloe vera (see page 164) and gotu kola (see page 166), while others such as sisal provide the starting materials for the manufacture of anti-inflammatory steroid drugs used for conditions including allergic dermatitis.

WASTE NOT WANT NOT

PLANT:
Agave sisalana Perrine
COMMON NAME(S):
sisal
FAMILY:
asparagus (Asparagaceae)

ACTIVE COMPOUND(S):
NATURALLY OCCURRING:
steroids (hecogenin)
MEDICINAL USES:
MAIN: pharmaceutical steroid
manufacture
OTHER: wounds
PARTS USED:
leaves

Hecogenin

ABOVE **After sisal (*Agave sisalana*) leaves have been processed to yield textile fibres, the waste left behind is a source of the steroidal compound hecogenin, which is used to synthesize pharmaceutical steroids.**

Sisal is a succulent plant native to Mexico but now cultivated in East Africa and Central America. It has long, thin, spine-tipped leaves that are a source of tough fibres used in the textile industry, often to make sturdy items such as sacks and ropes. But the plant holds another secret in its leaves: they provide the starting material for the production of around 5 per cent of global steroids for the pharmaceutical industry, making the use of this natural resource more efficient (see box).

Once the sisal fibres have been separated from the leaves by crushing, the leaf waste left behind is fermented for about a week and the ferment is then subjected to a purification process. The resulting product contains high levels of steroidal compounds, including hecogenin, which is chemically similar to steroids that occur in mammals. Hecogenin can be converted into corticosteroid drugs that suppress inflammation and the immune response, and so have a range of different pharmaceutical uses. Some corticosteroids are formulated as topical preparations, such as creams or ointments, and are applied to the skin to alleviate inflammatory conditions like eczema, dermatitis, and insect bites or stings.

A FLAVOUR OF PHARMACY

Many other plants also synthesize steroidal compounds, some of which have chemical structures that can be modified to produce useful steroidal pharmaceuticals. For example, the steroidal compound diosgenin from certain yams (*Dioscorea* spp.) in the yam family (Dioscoreaceae) has been used widely for the manufacture of contraceptive steroidal drugs (see page 186), but it can also be converted to corticosteroid drugs that have anti-inflammatory properties.

Yucca species, which are in the same plant family as sisal, and species of greenbrier (*Smilax* spp.) in the catbrier family (Smilacaceae), are sources of the steroidal compounds sarsasapogenin and smilagenin, which have also been used for pharmaceutical steroid manufacture. *Agave* and greenbrier species have another use in common – as drinks. Mexican sarsaparilla (*S. aristolochiifolia*) is one of the greenbrier species used to flavour the drink sarsaparilla, and blue or tequila agave (*Agave tequilana*) is used to make tequila.

ABOVE Mexican sarsaparilla (*Smilax aristolochiifolia*) is found from Mexico to Honduras. The stems of this woody climber can reach 5 m (16 ft) in length and bear heads of red fruit. Its roots are used in traditional medicine and to flavour sarsaparilla drink.

Global goals

The United Nations has 17 Sustainable Development Goals, also known as the Global Goals, which aim to end poverty and protect our planet. One of these goals is for responsible consumption and production, to reduce our ecological footprint. This includes recycling and reducing waste, and creating more efficient production and supply chains. More efficient use of our natural resources and discovering new ways to use plants sustainably can help contribute to this goal. Sisal is one example of how a plant may be utilized more efficiently, as once the fibres are removed for textile manufacture, the waste left over is used for the production of steroid-based drugs by the pharmaceutical industry.

LEFT: Sisal (*Agave sisalana*) was once cultivated solely for the production of sisal fibres (inset) for the textile industry but it now also supplies the pharmaceutical industry. Mature leaves, which may be up to 1.5 m (5 ft) long, are cut from the plant before being crushed and scraped to separate the fibres from the pulp. The fibres are then washed and hung up to dry in the sun.

A SOOTHING GEL

PLANT:

Aloe vera (L.) Burm.f.

COMMON NAME(S):

aloe vera

FAMILY:

asphodel or aloe
(Asphodelaceae)

ACTIVE COMPOUND(S):

NATURALLY OCCURRING:

polysaccharides

MEDICINAL USES:

MAIN: skin conditions

OTHER: constipation

PARTS USED:

leaves

C ut into an aloe vera leaf, and inside you will find a jelly-like substance that is often known as aloe vera gel. This is the parenchyma tissue of the leaf, which is mucilaginous and rich in complex carbohydrates (polysaccharides). Today, the gel is included in a wide range of products, from foods and food supplements to herbal medicines, cosmetics and household products.

Aloe vera has been used for centuries for its medicinal properties. It is considered native to Yemen and Saudi Arabia, although it has been introduced to many other regions of the world. The plant is reputed

ABOVE **Aloe vera (*Aloe vera*) is adapted to growing in arid conditions. It stores water in the central parenchyma tissue of its fleshy leaves and the thick, wax-coated skin (cuticle) reduces water loss through evaporation.**

to possess a plethora of medicinal benefits (see box) but is most widely known for its soothing properties. Scientific research suggests that enzymes in aloe vera gel are analgesic and counteract thermal damage, which could provide some explanation for its traditional use for burns. Other studies suggest the gel may be of some value in healing wounds and that it has anti-inflammatory properties. In clinical studies, some gel preparations applied to human skin or the mouth showed promising results for psoriasis, sunburn and mouth ulcers. However, further studies are needed to confirm any such benefits.

PORRIDGE PROTECTION

The oats we are familiar with in our breakfast porridge are the fruit (grain) of oat (*Avena sativa*) in the grass family (Poaceae), a descendant of wild oat (*Avena sterilis*), which originated in Iran and Iraq. Oat was domesticated about 3,000 years ago and has been cultivated in cooler regions of Europe, where it has become an important cereal crop. Oats have many reputed medicinal properties. They have been taken in the form of teas for nervous conditions, such as anxiety and insomnia, and are said to have restorative properties. Oat baths have also been a traditional approach for alleviating symptoms of rheumatism and gout.

Today, oats are finely ground into small particles and formulated into a colloid that is used in topical preparations such as creams and lotions for dry skin and related conditions. The grains are rich in polysaccharides, which comprise many units (molecules) of sugars linked together. When oat preparations are applied, the polysaccharides coat the skin, providing a type of protective layer that has a softening and soothing (emollient) or moisturizing effect. In addition, oats contain chemicals called avenanthramides, which have shown antioxidant and anti-inflammatory properties in laboratory studies and so might also soothe some skin conditions.

ABOVE Oat (*Avena sativa*) grains are used for their soothing properties in a range of skincare treatments and cleansing products.

Mascarene aloes

Aloe vera has many traditional uses. For example, in parts of South America, leaf exudates have been used for infections, for respiratory and digestive disorders, to alleviate pain and for injuries. However, it is for skin disorders such as inflammation, burns and wounds that aloe vera has become most famous as a traditional remedy.

Many other *Aloe* species are also known for their medicinal properties, including those from the Mascarene Islands in the Indian Ocean. For example, the Réunion aloe (*Aloe macra*), endemic to the island of that name, has been a traditional remedy for minor infections and boils, and used for healing skin. Laboratory studies have revealed that Mascarene aloes have antimicrobial activities, and that mazambron marron (*A. purpurea*), found on Mauritius, might also have some wound-healing properties. However, with the exception of aloe vera, the trade of all species of *Aloe* is regulated so that the survival of the plants in the wild is not threatened.

ABOVE Réunion aloe (*Aloe macra*) plant.

WORT FOR WOUNDS

PLANT:

Centella asiatica (L.) Urb.

COMMON NAME(S):

gotu kola, Asiatic pennywort

FAMILY:

carrot (Apiaceae)

ACTIVE COMPOUND(S):

NATURALLY OCCURRING:

triterpene saponins
(asiaticoside, madecassoside)

MEDICINAL USES:

MAIN: wounds

OTHER: nervous disorders

PARTS USED:

aerial parts

Gotu kola is a creeping perennial with kidney-shaped leaves and small pinkish-red flowers, and has a native range stretching from sub-Saharan Africa eastwards through tropical and subtropical regions to New Zealand and the southwest Pacific. The plant has a long history of use in traditional Indian medicine (see box). In herbal medicine, extracts of the aerial parts have been of much interest to aid the healing of wounds, including burns.

Laboratory studies have shown that extracts of the aerial parts, and some of the triterpenoid glycoside constituents (asiaticoside and madecassoside) and triterpenoid constituents (asiatic acid and madecassic acid), stimulate the production of collagen, a protein in the skin that has an important role in wound healing.

Extracts from the plant and these constituents also have anti-inflammatory properties, which might help explain their traditional use for inflamed skin conditions such as burns. One clinical study claimed that a cream preparation of gotu kola applied to human skin provided some improvement in the appearance of scars. A separate clinical trial suggested that gotu kola improved wound healing in people with diabetes when taken orally in the form of capsules. More studies are needed to evaluate further the usefulness of gotu kola preparations for skin conditions.

FRUIT FIXERS

Another plant-derived substance used for burns and inflammation is bromelain, an enzyme in the fruit of the pineapple (*Ananas comosus*) in the bromeliad family (Bromeliaceae). The enzyme papain, found in the fruit of papaya (*Carica papaya*) in the papaya family (Caricaceae), is also used for similar purposes. These plant enzymes are known as skin-debriding agents because they contribute to the breakdown of protein in the skin and so help to remove damaged skin cells, allowing the healthy tissue to heal.

BUMPS AND BRUISES

Arnica flower heads (capitula) have long been applied to the skin as a traditional remedy for bruises and other contusions, as well as for aches and sprains, and extracts are also said to aid wound healing. The flower heads contain sesquiterpene lactones, such as helenalin, which are considered to be the principal anti-inflammatory constituents. Arnica has been tested in clinical studies in humans with conditions such as bruises and osteoarthritis, and also poor circulation that causes a feeling of heaviness in the legs. These studies suggest that applying arnica preparations to the skin might alleviate some symptoms, but additional scientific research is needed to investigate the potential medicinal properties of arnica more extensively.

FAR LEFT Gotu kola or Asiatic pennywort (*Centella asiatica*) is a small, trailing evergreen perennial with kidney-shaped leaves and inconspicuous flowers, and grows in wet conditions. It is used in many systems of traditional medicine, including Ayurvedic, and should not be confused with another medicinal plant, *Bacopa monnieri* in the speedwell family (Plantaginaceae), with which it shares common names including brahmi and Indian pennywort.

LEFT Arnica (*Arnica montana*) is a perennial herb native to Europe, including Scandinavia, and Greenland, where it grows in mountain grassland and open woods. Bright yellow flower heads are produced on slender stems above a basal rosette of leaves.

ABOVE Papaya (*Carica papaya*) is a slender tree whose edible fruit turn orange or yellow when ripe. It is native to a region extending from Mexico to Venezuela and is widely cultivated in tropical countries.

Penny for your thoughts

Gotu kola is also known as the Asiatic or Indian pennywort, and has been used in Ayurvedic medicine for its reputed ability to restore memory and for longevity. When combined with other herbs, it is reputed to delay symptoms of ageing and prevent dementia. In traditional Chinese medicine, the plant has been used to combat physical and mental exhaustion. Native Americans combined it with other plants as a traditional remedy to assist memory. Laboratory studies have shown that extracts of the aerial parts may help protect nerve cells. In some preliminary studies in humans, gotu kola preparations appeared to improve memory and reduce anxiety, although more clinical studies are needed to confirm any such benefits.

Psoriasis

Psoriasis is an inflammatory skin condition, certain forms of which are treated with dithranol preparations. The production of dithranol was inspired by a chemical that occurs in Goa powder, obtained from a Brazilian tree and also found in the roots of yellow dock (*Rumex crispus*) in the knotweed family (Polygonaceae). Greater ammi and other plants contain psoralen chemicals, which in certain controlled circumstances may also be used therapeutically for conditions such as psoriasis.

LIGHT RELIEF

PLANT:

Ammi majus L.

COMMON NAME(S):

greater ammi, bishop's weed, lady's lace

FAMILY:

carrot (Apiaceae)

ACTIVE COMPOUND(S):

NATURALLY OCCURRING:

psoralens (methoxsalen)

MEDICINAL USES:

MAIN: psoriasis, vitiligo

OTHER: respiratory conditions

PARTS USED:

fruit

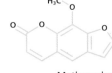

Methoxsalen

LEFT **Greater ammi (*Ammi majus*)** fruit contain psoralens, including methoxsalen, which increases the skin's sensitivity to ultraviolet radiation. Preparations of methoxsalen are used in photochemotherapy for vitiligo and certain forms of psoriasis.

Greater ammi fruit, often referred to as 'seeds', contain chemicals called psoralens. When preparations incorporating psoralens are applied to the skin, they can cause a reaction on exposure to sunlight. This action would usually be considered an adverse effect, but because psoralens increase the skin's sensitivity to ultraviolet A (UVA) wavelengths of light, this specificity can be used as the basis of psoralen-UVA (PUVA) photochemotherapy. For example, the chemical methoxsalen (8-methoxypsoralen) in the fruit, when applied to the skin or taken orally in conjunction with exposure to UVA, binds to the DNA in skin cells. This inhibits cell division, causing cell injury, and as the skin recovers, the outer layer (stratum corneum) can thicken and the production of the skin pigment melanin may increase. These actions explain why psoralens combined with UVA light are specialist treatments for conditions such as psoriasis and vitiligo. Similar types of chemicals also occur in the roots of chhataya (*Heracleum candicans*), another member of the carrot family, and the fruits of the fountain bush or bakuchi (*Cullen corylifolium*, syn. *Psoralea corylifolia*; see page 31) in the legume family (see box). Chhataya cultivated in India is now one of the main sources of psoralens for use in PUVA.

LEFT **Greater ammi (*Ammi majus*) is native to the countries of Europe and North Africa bordering the Mediterranean, and also to the Middle East. It has divided leaves and flat-topped heads of small white flowers.**

Greater ammi has been used traditionally in different practices of medicine. For example, in Egypt the fruit or leaves were used for various skin conditions. In traditional Chinese medicine, the plant has been used for heart and respiratory conditions, and in Saudi traditional medicine it was used for urinary problems.

The fountain bush, which also contains psoralens, has been a traditional Chinese medicine for centuries, famous for its reputed tonic properties. The plant has also been used in Ayurvedic and Unani systems of medicine, as a remedy for skin conditions such as boils, eczema and itching.

LEFT The roots of yellow or curly dock (*Rumex crispus*), a perennial herb growing to 2–2.5 m (6.5–8 ft) tall, are used medicinally. The species is native to North Africa and across Europe to far east and northern Siberia, but widely naturalized elsewhere.

BELOW Pods (fruit) of the Brazilian araroba tree (*Vataireopsis araroba*). The tree's wood is the source of Goa powder, which was used traditionally for psoriasis.

A TREE TREATMENT

Goa powder, also referred to as araroba powder, is obtained from the wood of the araroba tree, native to Brazil. The Portuguese exported it from Brazil to their colony of Goa in India, and it became a traditional remedy for psoriasis there and also in China. It was prepared by mixing the powder with water or lime juice or vinegar to form a paste, which was then applied to the skin. A London dermatologist, Balmanno Squire (1836–1908), recorded this use of Goa powder and its main active component, the anthrone derivative chrysarobin, in his 1878 treatise. During the First World War, trade with Brazil was disrupted and so supplies of chrysarobin became depleted. To develop new psoriasis preparations to replace chrysarobin, substitutes based on the chrysarobin chemical structure were synthesized. One of these was dithranol, also known as anthralin. Dithranol paste was subsequently introduced for certain forms of psoriasis and topical preparations are still widely prescribed today. Dithranol is thought to interfere with skin cell proliferation and inflammation, and mediates these effects when it is applied to the skin.

Mastering microbes

Skin may become infected with microbes such as bacteria and fungi, causing unpleasant symptoms. Some plants, including tea tree, have antimicrobial properties and so can be used as antiseptics. Certain plant constituents act in other ways to reduce the symptoms caused by microbial infections of the skin.

ANTISEPTIC TREE

PLANT:

Melaleuca alternifolia
(Maiden & Betche) Cheel

COMMON NAME(S):

tea tree

FAMILY:

myrtle (Myrtaceae)

ACTIVE COMPOUND(S):

NATURALLY OCCURRING:

monoterpenes (terpinen-4-ol)

MEDICINAL USES:

MAIN: skin infections

OTHER: cold and flu symptoms

PARTS USED:

leaves and branchlets

RIGHT **Tea tree (*Melaleuca alternifolia*) essential oil contains about a hundred chemical constituents, including terpinen-4-ol, which has shown antimicrobial and other biological activities in scientific studies.**

Terpinen-4-ol

The leaves and branchlets of tea tree, a small tree native to Australia, have a distinctive odour thanks to the essential oil they produce, which is rich in fragrant terpene chemicals, especially monoterpenes. Today, tea tree oil is widely available for its antiseptic properties, often in the form of cosmetic products such as washes, lotions and shampoos.

Scientific research appears to provide some support for the use of tea tree oil against skin infections caused by bacteria and fungi. Numerous laboratory studies show that the oil, and one of its principal active constituents, terpinen-4-ol, have an action against bacteria that colonize the skin in acne (for example, *Cutibacterium acnes*). Tea tree oil has been tested in clinical studies, diluted in formulations such as creams and applied to the skin, to assess any usefulness in treating acne. Some of these studies in humans suggest

BELOW **Flowering branch of tea tree (*Melaleuca alternifolia*), also known as narrow-leaf paperbark.**

that such preparations might help reduce the severity of symptoms in acne. This may be because the oil has both antibacterial and anti-inflammatory properties.

Tea tree oil has been of much interest for the antibacterial effects it has against methicillin-resistant *Staphylococcus aureus* (MRSA), a strain of bacteria that is becoming increasingly difficult to treat using conventional antibiotics as it has developed strategies to resist their effects. In other scientific studies, tea tree oil has shown activity against different types of fungi that can cause skin infections, including athlete's foot and vaginal thrush. Small clinical studies in humans also suggest the oil might be useful against these and some other fungal infections of the skin and nails, and it appears to reduce symptoms of dandruff when used in the form of a shampoo. In addition, tea tree oil is active against some viruses such as *Herpes simplex*, which causes cold sores, where terpinen-4-ol is again observed to be the main active constituent.

WONDER WILLOW

In acne, the production of sebum – the oily secretion from the skin's sebaceous glands – is increased. This causes pores to become blocked with both the sebum itself and skin cells, and bacterial colonies then trigger inflammation, leading to pimples. Salicylic acid, which occurs in plants such as the willow and meadowsweet

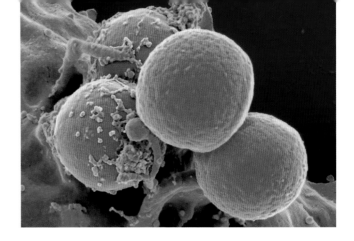

ABOVE Coloured scanning electron micrograph of methicillin-resistant *Staphylococcus aureus* (MRSA) bacteria.

(*Filipendula ulmaria*) (see page 142), has an anti-inflammatory action and breaks down keratin in the surface layer of the skin (it is keratolytic), which helps to unblock pores. It is therefore included in many topical preparations for acne.

Salicylic acid has antifungal as well as keratolytic properties, so it is included in some preparations for scaling skin conditions such as dandruff and psoriasis. Salicylates are used medicinally for a number of conditions in addition to skin problems (see also page 173), ranging from heart disease (see page 38) to pain (see page 142). It could be said that the plants from which they are derived, such as the willow, really have been a wonder in the discovery of medicines.

The original antiseptic

Tea tree leaves and the oil obtained from them were traditionally used by Aboriginal Australians as topical remedies for skin infections, insect bites, wounds and bruises. Interest in producing tea tree oil commercially began in the 1920s, and legend has it that Australian soldiers were supplied with the oil in their military kits during the Second World War. After antibiotics such as the penicillins were introduced in the 1940s, use of tea tree oil as an antiseptic declined. However, it remains a popular ingredient in many cosmetic and aromatherapy products.

RIGHT Tea tree (*Melaleuca alternifolia*) oil is distilled from the tree's small leafy branches.

Warts and all

Warts and verrucas are skin infections caused by the human papilloma virus. Current treatments involve applying medicinal substances that destroy the infected skin cells, and include those derived from mayapple rhizome. Sinecatechins are chemicals from green tea (*Camellia sinensis*) in the tea family (Theaceae), and are also applied to the skin as a wart treatment. They are considered to have multiple actions against viral skin infections.

RESIN TO THE RESCUE

PLANT:
Podophyllum peltatum L.

COMMON NAME(S):
mayapple, American mandrake

FAMILY:
barberry (Berberidaceae)

ACTIVE COMPOUND(S):

NATURALLY OCCURRING:
lignans (podophyllotoxin)

MEDICINAL USES:

MAIN: warts

OTHER: anti-tumour

PARTS USED:
rhizomes and roots

The mayapple plant, also known as American mandrake, is native to southeast Canada and central and east United States. Rhizome preparations were included in the materia medica list of the first United States Pharmacopoeia in 1820, although their medicinal uses date back even further (see box). The rhizomes produce a resin known as podophyllin, which contains lignan constituents such as podophyllotoxin (see page 202). When applied to warts, both the resin and podophyllotoxin are toxic to the skin cells infected with the wart-causing virus, and

ABOVE Himalayan mayapple (*Podophyllum hexandrum*) fruit can be up to 7 cm (2½ in) long and are used medicinally in China to regulate menstruation.

LEFT AND RIGHT Mayapple (*Podophyllum peltatum*) is a perennial herb with creeping rhizomes that are used medicinally. Stems with one or two large, lobed leaves are produced at intervals, and white (or occasionally pink) flowers arise singly from the leaf joints.

so these are destroyed. Podophyllotoxin may also modulate immune function in the area of the skin to which it is applied, to help combat the virus. Laboratory studies suggest this lignan may have anti-inflammatory and antiviral properties. These actions explain why both podophyllin resin and podophyllotoxin are used as topical preparations to treat certain types of warts.

The rhizome of another plant, the Himalayan or Indian mayapple (*Podophyllum hexandrum*), contains more resin compared to the mayapple, and also much higher amounts of podophyllotoxin. For this reason, it is used as a source of the resin and podophyllotoxin for the pharmaceutical industry to produce topical medicines to treat warts. It is also a source of podophyllotoxin used in the manufacture of important anticancer drugs.

NOT JUST YOUR CUP OF TEA

Green tea leaves contain many different catechins, complex chemicals that are known to bind to different proteins, including those involved in generating inflammatory substances. Sinecatechins are a mixture of the green tea catechins, the principal of which is epigallocatechin gallate. Sinecatchins have been prepared as an ointment and applied to the skin to clear particular types of warts, as they are claimed to have antiviral and immune-stimulating effects.

Salicylates, originally derived from willow bark and meadowsweet, include salicylic acid. This is another topical treatment applied to the skin to treat warts and verrucas, as it is keratolytic, helping to break down infected skin cells (see also pages 142 and 171).

A potent purgative

Native Americans used mayapple rhizome as a traditional remedy for various conditions. Preparations were taken in small doses to kill parasitic worms and induce vomiting, and as a purgative. The purgative action is now known to be largely due to chemicals called peltatins present in the rhizome resin. Traditionally, the rhizome was also reputed to act as a liver tonic. However, it is known to be toxic when ingested and may cause symptoms of paralysis. This toxicity has been exploited for other applications, as preparations of the mayapple were traditionally applied to potato crops as an insecticide.

MALESIA

Malesia is the region of Southeast Asia stretching from Malaysia in the west, through the 17,000 islands that make up Indonesia to the south and east, and to the Philippines in the north. It is predominantly tropical, with cooler climatic zones on the highest mountains, and is floristically rich, with an estimated 28,000 plant species in Indonesia alone. Although many native plants are relied on to provide local medicines, the traditional systems are little known outside the region.

Spice Islands

Evidence for the use of cloves (*Syzygium aromaticum*; see page 77) from the myrtle family (Myrtaceae) in China in the third century BCE, as well as their description by the Roman naturalist Pliny the Elder (23–79 CE), gives an indication of the extent of early trade between the islands of Malesia and the rest of the world. The desire for exotic spices in Europe, which started during the Middle Ages, fuelled the age of discovery in the subsequent centuries and brought Europeans to Malesia as explorers, traders and conquerors. In particular, they were interested in the Maluku Islands (Moluccas), also known as the Spice Islands and now part of Indonesia. The archipelago was the sole source of cloves, as well as nutmeg and mace, which both come from the fruit of the nutmeg tree (*Myristica fragrans*) in the nutmeg family (Myristaceae; see box).

ABOVE **The nutmeg tree (*Myristica fragrans*) produces the spice of the same name, which is the seed kernel, while mace is the aril, a netted, slightly fleshy coating to the seed that is red when fresh.**

The Dutch controlled the area that was to become Indonesia, and therefore the trade in these spices, for almost 300 years. During the Napoleonic Wars at the turn of the nineteenth century, the British were able to infiltrate the islands where nutmeg grew and obtain seedlings, which they used to start plantations elsewhere and so broke the dominance of the Dutch in the spice trade. It was the Second World War, during which the region was occupied by the Japanese, that eventually led to independence for Indonesia from the Dutch in 1949, and of Malaysia from the British in 1957.

ABOVE **Jamu medicines for sale in a traditional market, Yogyakarta, Indonesia. Jamu is a traditional Malesian system of medicine, in which remedies are usually taken in the form of freshly made drinks.**

Jamu

The Europeans still relied on a system of medicine dominated by humoral explanations and herbal remedies when they arrived in Indonesia in the sixteenth century. This was comparable to the traditional medicine they encountered in the region, which is used to treat conditions that indicate an imbalance between the poles of hot and cold (*panas* and *sejuk*, respectively, in Malay), both of which are essential to health. The predominant system of medicine in Malesia is Jamu. According to tradition, it originated in the principal courts of Java, located in Surakarta and Yogyakarta. It is now widely practised in both Indonesia and Malaysia by people from all classes of society.

Jamu uses an estimated 6,000 plant species, including some that indicate the influence of Chinese, Indian and Arabic systems of medicine. Medicines can contain up to 40 plant species and are prepared by women in the home. They are used fresh, both to treat illness and as tonics to strengthen and refresh. Recipes are passed down from mother to daughter and are often kept secret. Some women supplement their income by selling medicines to neighbours, and in cities street vendors hawk remedies. In more recent years, preparations in the form of powders, pills and pastes have become available from an increasing number of manufacturers.

FRAGRANCE AND FLAVOUR

In medieval Europe, fragrance had an emotional and spiritual value, and aromatic plants were valued in food, as medicines and as perfumes. The more exotic and expensive they were, the more prestige was gained from their use. Aromatic substances from Malesia included nutmeg and mace. In the Maluku Islands today, nutmeg oil is rubbed on the abdomen for the relief of stomach ache and on the temples for headache. In Java, both nutmeg and mace are used to make a calming drink for those suffering from insomnia or stress.

Ylang ylang (*Cananga odorata*) in the soursop family (Annonaceae) is widely distributed throughout Malesia, and is also found in Queensland, Australia. The essential oil from the flowers of this tree is used in perfumes and aromatherapy. Benzoin, the resin from the gum Benjamin tree (*Styrax benzoin*) in the storax family (Styracaceae), a species thought to be native to Sumatra, is now used medicinally as an inhalation to relieve catarrh, and as an antiseptic when applied to the skin.

ABOVE **Leafy ylang ylang (*Cananga odorata*) branch bearing a sweetly scented flower.**

Bug busters

Some parasites can cause infestations in humans, including scabies mites and head lice (*Pediculus humanus capitis*). Peru balsam is obtained from the bark of a tree and was once used as a traditional remedy for scabies; today, a chemical that occurs in the balsam is applied to the skin for the same purpose. Certain plant oils, including tea tree oil, have also been studied for their usefulness against these parasites.

A MIGHTY BALSAM

PLANT:
Myroxylon balsamum (L.) Harms (syn. *Myroxylon pereirae* (Royle) Klotzsch)

COMMON NAME(S):
Peru balsam, tolu balsam

FAMILY:
legume (Fabaceae)

ACTIVE COMPOUND(S):
NATURALLY OCCURRING:
balsamic esters (benzyl benzoate)

MEDICINAL USES:
MAIN: skin parasites
OTHER: haemorrhoids

PARTS USED:
trunk and bark

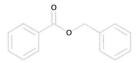

Benzyl benzoate

ABOVE **Peru balsam is a viscous liquid considered to have mild antiseptic properties. It contains benzyl benzoate, a compound that is used in some preparations for scabies.**

BELOW **Peru balsam (*Myroxylon balsamum*) is a slow-growing tree found in the lowland tropical forests of Central America and northern South America, where it can reach 45 m (150 ft) in height.**

Peru balsam trees are native to Mexico, Peru and other parts of tropical South America. The balsam is extracted from the trunks and methods used to prepare it date back to the sixteenth century. Peru balsam itself is a viscous brown liquid that is obtained by scorching the living bark with fire and wounding it to cause damage, which stimulates production of the balsam. The following week, strips of bark are removed and then crushed, boiled and pressed to retrieve the balsam. The exposed trunks are also a source of balsam that can be soaked up over a period of days with fabric. The trunk is then covered with fabric a second time to collect a third exudation. On each occasion, the fabric is gently boiled in water and pressed, during which time the balsam sinks and the water is poured away to leave it behind. Balsam from the three extractions is then mixed in specific proportions before being boiled to remove residual water. The final product is Peru balsam.

Peru balsam contains a resin and balsamic esters, including benzyl benzoate, which is toxic to some parasites that can affect humans. When used as a pharmaceutical, benzyl benzoate alone is formulated in emulsions or similar liquid preparations that are applied to the skin to treat the mites that can cause scabies, and also sometimes head lice.

FRAGRANT BUT FIERCE

Some plants in the mint family (Lamiaceae) produce essential oils that have been of interest for their use against skin parasites such as scabies mites and head lice. One such plant is thyme (*Thymus vulgaris*), which we often use for its distinctive aroma and flavour when cooking. Thyme essential oil has been of interest for

Secret behind the balsam

Original uses of Peru balsam involved diluting it in the oils of plants such as castor oil (*Ricinus communis*) in the spurge family (Euphorbiaceae), and applying such preparations to alleviate bedsores and skin ulcers, and sometimes itchy skin. It was also traditionally applied for other skin conditions such as wounds, burns, frostbite and scabies mites, and was once used for rheumatic complaints and as an ingredient of some inhalant preparations for the relief of nasal congestion. Today, Peru balsam is still included in some preparations that are applied to relieve symptoms of haemorrhoids.

use against both scabies mites and head lice, but more scientific evidence is needed to support this use. Other plant essential oils that have been studied for their usefulness against head lice include oil from lavender (*Lavandula angustifolia*) in the mint family, and tea tree oil. When applied in combination, tea tree and lavender oil preparations have shown some promising results.

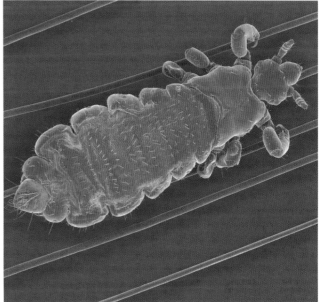

ABOVE Coloured scanning electron micrograph of a head louse (*Pediculus humanus capitis*) on human hair. Head lice are parasites that live near the scalp and feed on their host's blood.

LEFT Thyme (*Thymus vulgaris*) is a low-growing, aromatic herb native to Spain, France and Italy.

The eyes have it

In glaucoma, the channels that allow fluid to drain from the eye become blocked, causing pressure to build up, which can affect vision. Medicines that relieve this pressure include eye drops containing pilocarpine, from the shrub Maranham jaborandi, and physostigmine from Calabar beans. For certain other eye conditions, atropine drops are used, derived from some plants in the potato family, including belladonna and mandrake (*Mandragora officinarum*) (see page 180).

RELIEVING THE PRESSURE

PLANT:

Pilocarpus microphyllus Stapf ex Wardlew.

COMMON NAME(S):

Maranham jaborandi

FAMILY:

citrus (Rutaceae)

ACTIVE COMPOUND(S):

NATURALLY OCCURRING:

imidazole alkaloids (pilocarpine)

MEDICINAL USES:

MAIN: glaucoma

OTHER: skin conditions

PARTS USED:

leaves

Pilocarpine

LEFT The imidazole alkaloid pilocarpine occurs in the leaves of Maranham jaborandi (*Pilocarpus microphyllus*) at concentrations of 0.7–0.8 per cent. It is formulated in some eye preparations and used pharmaceutically for glaucoma.

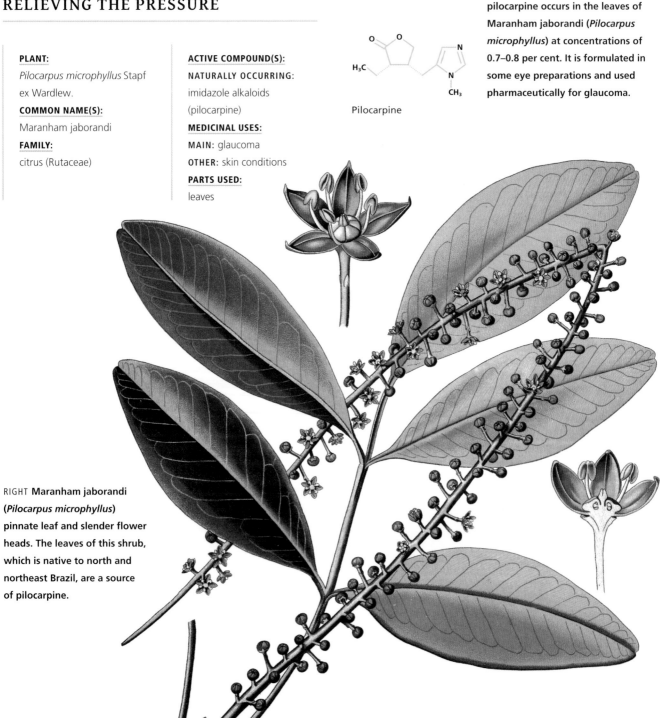

RIGHT Maranham jaborandi (*Pilocarpus microphyllus*) pinnate leaf and slender flower heads. The leaves of this shrub, which is native to north and northeast Brazil, are a source of pilocarpine.

Maranham jaborandi and some other plants in the same genus that are also native to Brazil, such as Paraguay or Pernambuco jaborandi (*Pilocarpus jaborandi*), contain the alkaloid pilocarpine. Maranham jaborandi leaves contain higher levels than other jaborandi species, and so are used as a source of this pharmaceutically important chemical.

In the condition glaucoma, pressure in the eyes is increased and may result in damage to nerves in the eyes, including the optic nerve. Pilocarpine mimics the action of the neurotransmitter acetylcholine and acts on receptors in the nervous system called muscarinic receptors. When applied to the eyes in the form of eye drops, it influences the muscles in the eyes and causes the pupils to constrict (miosis). This action facilitates the drainage of fluid from the eyes, reducing the pressure. Pilocarpine is also sometimes used in the form of an oral medicine to alleviate dry eyes or a dry mouth caused by certain conditions, including some radiotherapy treatments.

CALABAR BEANS

Calabar beans contain the indole alkaloid physostigmine, which inspired the development of a drug to treat dementia symptoms (see page 72). Physostigmine inhibits acetylcholinesterase, and so prevents this enzyme from breaking down the neurotransmitter acetylcholine. As a result, the action of acetylcholine is prolonged in the nervous system. This means that the effects of physostigmine are similar to those of pilocarpine from jaborandi plants, even though they are acting in different ways. For this reason, physostigmine has also been applied as eye drops to reduce the pressure in the eyes that occurs in glaucoma.

Hair-raising uses

The first record of jaborandi plants was in 1587. Historically, jaborandi leaf infusions were part of shamanic rituals in the Amazon region, and were considered to combat fevers and mouth inflammation, and used as an antidote to poisons. Jaborandi leaf preparations were also used traditionally to alleviate itchy skin conditions such as psoriasis and were reputed to combat hair loss. They were also traditional remedies for syphilis, catarrh and oedema.

In 1873, the Brazilian-born physician Sinfrônio Olímpio César Coutinho (1832–1887) first introduced Europeans to the medicinal properties of the native jaborandi plants. The name jaborandi is derived from the Tupí–Guaraní word *ya-mbor-endi*, meaning 'what causes slobbering'. Indeed, infusions of the leaves were known to stimulate sweat and saliva production. This can be explained by the presence of the alkaloid pilocarpine, which was first isolated from the plant in 1875, and was later revealed to stimulate the cholinergic nerves to produce salivation and other effects.

BELOW Calabar bean (*Physostigma venenosum*), showing the hilum, a scar on the seed coat from which it was originally attached to the ovary placenta by a stalk (funicle).

BEAUTY IS IN THE EYE OF THE BEHOLDER

PLANT:
Atropa bella-donna L.
COMMON NAME(S):
belladonna, deadly nightshade
FAMILY:
potato (Solanaceae)

ACTIVE COMPOUND(S):
NATURALLY OCCURRING:
tropane alkaloids (hyoscyamine)
MEDICINAL USES:
MAIN: eye conditions
OTHER: rheumatism
PARTS USED:
leaves, roots

The name belladonna means 'beautiful woman' in Italian, and originates from the traditional use of the juice from the plant's dark, shiny berries. In Italy, women would drop this juice into their eyes to dilate their pupils, making them appear more attractive.

Belladonna has a native range extending across west and central Europe to the Mediterranean and northern Iran. Its traditional adoption as an early beauty product has evolved into the use of some of the plant's chemical constituents in modern medicine. The main alkaloid constituent of belladonna leaves and roots is (–)-hyoscyamine (see page 156); this can change into (+)-hyoscyamine, a much less medicinally active form. The mixture of the two forms is known as atropine, which has been developed as a pharmaceutical drug for different applications.

Atropine blocks the action of the neurotransmitter acetylcholine, so when applied in the form of eye drops it causes the muscles in the eye to relax, dilating the pupil. This is useful when examining the eyes, so ophthalmologists utilize atropine drops for this reason. In certain eye conditions such as uveitis, which involves inflammation in part of the eye, atropine drops are sometimes used to help relieve symptoms.

ABOVE AND LEFT **Belladonna or deadly nightshade (*Atropa bella-donna*) is a herbaceous perennial that can reach 1.5–2 m (5–6.5 ft) in height. Bell-shaped flowers appear singly, followed by juicy black fruit that have caused poisoning when mistaken for those of an edible plant.**

Atropine has many other uses in medicine, from relieving muscle spasms in the gut to managing some heart conditions. It is also used as a premedication to dry up bodily secretions such as saliva before surgery, and as an antidote following poisoning with some types of pesticides or nerve agents. Furthermore, the chemical structure of atropine has been the basis for the design of new drugs that are used to alleviate symptoms in Parkinson's disease. Atropine shares some of these applications with tropane alkaloids from other members of the potato family (see pages 86, 91 and 156).

ABOVE Mandrake (*Mandragora officinarum*) is a herbaceous perennial native to north Italy and the northwest Balkans. It bears a rosette of large leaves, and fruit to 4 cm (1½ in) in diameter, above a thick tap root.

MANDRAKE MYTHS AND MAGIC

Mandrake is also in the potato family and, like belladonna, is a source of atropine. Throughout history, the plant has been shrouded in magic, myth and superstition. In the Middle Ages, sorcerers believed the root could be half vegetable and half human – according to legend, the root would shriek when pulled from the ground and anyone who heard this would suffer madness or death. To prevent this, herbalists would tie a dog to the plant, move beyond hearing range of the deathly shriek and then call the dog, so that it pulled the root up and its master remained safe. Mandrake has also become famous in more recent times, being referred to in books about a young wizard called Harry Potter.

A witch's potion

The first records of belladonna use date back to the early sixteenth century, and it has a long-standing reputation for both medicinal and poisonous properties. Its use is steeped in folklore and superstition, with records from around 500 years ago suggesting witches used belladonna to anoint their bodies and that this created a sensation of flying. In the Scottish Highlands, it was believed that people who used belladonna would see ghosts.

Belladonna leaf preparations were traditionally applied to relieve rheumatic pains and chilblains, while decoctions were taken to aid circulation and counteract diarrhoea. Belladonna stems were also cut and joined together to make a necklace that was then worn by children in the belief it would prevent teething pain.

LEFT Witch depicted flying on a broom.

CHAPTER 9

INFLUENCING THE REPRODUCTIVE SYSTEM

The reproductive system is the most obvious way in which the male and female sexes differ. For this reason, most of the conditions in this chapter apply to only one sex. Some plants and plant compounds can benefit conditions associated with the menstrual cycle, menopause and impotence, and female contraceptive pills were revolutionized by a plant compound, but the quest for the male pill continues.

Chasteberry (*Vitex agnus-castus*)

NATURAL COMPOUNDS FOR THE REPRODUCTIVE SYSTEM

The ability to control fertility and ameliorate symptoms associated with the menstrual cycle and menopause have benefited millions of women worldwide, and plants have been at the forefront of this revolution. A plant compound has also showed promise as a contraceptive for men, and others have been of interest for impotence.

ABOVE Maca (*Lepidium meyenii*) is a low-growing biennial with divided leaves that hug the ground and central heads of small self-fertile flowers (for roots, see the image opposite).

FERTILITY CONTROL

The discovery that plants produce compounds that can act as starting material for the production of human sex hormones opened up the possibility of affordable pharmaceuticals. Yams (*Dioscorea* spp.; see page 186) in the yam family (Dioscoreaceae) played an important part in this, particularly the Mexican species cabeza de negro (*D. mexicana*) and barbasco (*D. composita*). They produce diosgenin, a sapogenin, and resulting drugs revolutionized birth control for women, as well as providing treatments for the menopause and inflammation. Cotton plants (*Gossypium* spp.; see page 188) in the mallow family (Malvaceae) might be an unlikely source of a contraceptive for men, but consuming crude cottonseed oil was found to cause male infertility. A safe pharmaceutical has yet to be developed from the active compound, (–)-gossypol.

MONTHLY CYCLE

With female fertility comes a monthly cycle of changes that prepare the body for pregnancy and then reset it if the released egg (ovum) is not fertilized. Many women suffer from premenstrual syndrome (PMS), caused by imbalances in the associated hormonal changes, and herbal preparations and dietary supplements may offer some relief. Chasteberry (*Vitex agnus-castus*; see page 192) in the mint family (Lamiaceae) was a symbol for chastity and the fruit were thought to suppress male sexual desire. The seeds of several plants are a source of *gamma*-linolenic acid (GLA), which is a precursor to important compounds in the body. Most research has looked at seed oil from evening primrose (*Oenothera biennis*; see page 193) in the evening primrose family (Onagraceae) as a source of GLA, but the seed oil from borage or starflower (*Borago officinalis*) in the borage family (Boraginaceae) contains higher amounts.

Most women stop ovulating by their early 50s, but the transition from fertile to infertile, known as the menopause, can be accompanied by symptoms that affect quality of life. Black cohosh (*Actaea racemosa*; see page 194) in the buttercup family (Ranunculaceae) is a herbal preparation that was traditionally used by Native American women for various conditions associated with the menstrual cycle and is now widely used for the menopause. Plant compounds with oestorgenic activity are also of interest, including those found in members of the legume family (Fabaceae; see page 195).

STIMULATING IDEAS

Numerous plants worldwide are used traditionally as aphrodisiacs to stimulate sexual desire. Far fewer are used for impotence in men. The bark of yohimbe (*Corynanthe johimbe*; see page 196) in the coffee family (Rubiaceae) is used ritualistically in regions of west Central Africa, and although there is some evidence to support its use for impotence, the risks are considered to outweigh the benefits. Barrenwort (*Epimedium* spp.; see page 197) in the barberry family (Berberidaceae) captures the imagination with its alternative name of horny goat weed; it has been used for several thousand years in traditional Chinese medicine. And the roots of maca or Peruvian ginseng (*Lepidium meyenii*) in the cabbage family (Brassicaceae), from the high altitudes of Peru, are said to have been used by Inca warriors to give them strength before battle and are still eaten regularly as a vegetable. Further research is required to determine if this species is of benefit for impotence.

ABOVE **Oil from the seeds of borage (*Borago officinalis*) is a source of *gamma*-linolenic acid and should not contain the toxic pyrrolizidine alkaloids that are present in other parts of the plant.**

ABOVE **Maca (*Lepidium meyenii*) roots, which grow to 8 cm (3 in) wide, are eaten as a vegetable and are of interest for impotence.**

LEFT **Cabeza de negro or Mexican yam (*Dioscorea mexicana*) tubers can reach 90 cm (35 in) in diameter and were a source of diosgenin during the development of semi-synthetic human hormones.**

Family planning

Historically, childbirth has been a significant risk to the well-being of women, and in areas with limited healthcare this is still the case. The introduction of reliable birth control options that enabled people to plan when to have children has transformed lives. These options include 'the pill' for women, which was developed thanks to a compound found in yams. Meanwhile, the search for a contraceptive pill for men continues (see page 188).

SEX HORMONE SOURCES

PLANT:

Dioscorea mexicana Scheidw. and *Dioscorea composita* Hemsl.

COMMON NAME(S):

cabeza de negro; barbasco

FAMILY:

yam (Dioscoreaceae)

ACTIVE COMPOUND(S):

NATURALLY OCCURRING:

steroidal saponin (dioscin); sapogenin [aglycone] (diosgenin)

SEMI-SYNTHESIZED:

norethisterone, progesterone, testosterone and others; cortisone

MEDICINAL USES:

MAIN: oral contraceptive

OTHER: premenstrual syndrome, menopause; inflammatory diseases

PARTS USED:

tuberous roots

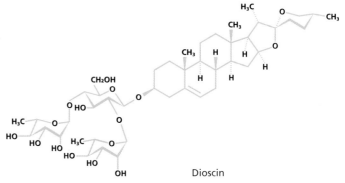

Dioscin

ABOVE **The most abundant steroidal saponin present in wild yam (*Dioscorea villosa*) is dioscin. In the manufacture of steroidal drugs, the sugar portion of the compound is removed by acid hydrolysis or other processes to form diosgenin, an aglycone or sapogenin.**

ABOVE **Barbasco (*Dioscorea composita*), like many other species of yam, is a perennial climber with heart-shaped leaves.**

In the 1930s, the structures of several sex hormones – including progesterone and testosterone – were determined and their medical applications explored. However, the high cost of producing these from animal sources such as the urine of pregnant horses meant that they had only limited availability. In 1938, the American chemist Russell Earl Marker (1902–1995) devised a series of chemical reactions that could produce progesterone using, as its starting material, sarsasapogenin, a steroidal sapogenin from the roots of Mexican sarsaparilla (*Smilax aristolochiifolia*; see page 163) in the catbrier family (Smilacaceae). This was also too costly to produce commercially, so Marker looked for an alternative raw material.

Marker found that diosgenin, which had previously been identified by Japanese chemists as a sapogenin from tokoro (*Dioscorea tokoro*), a species of yam, was more similar in structure to progesterone and conversion would therefore involve fewer steps. This was confirmed using the roots of wild yam (*D. villosa*), a North American species, and the process came to be known as the Marker degradation. Marker went on to look at more than 400 plants, collected for him by botanists, in his quest for an inexpensive source of diosgenin, but it was an image in a book that gave him the answer. A species of Mexican yam known as cabeza de negro (*D. mexicana*) had enormous root tubers and did indeed yield good levels of diosgenin.

MEXICAN YAMS

Unable to interest any American pharmaceutical companies in taking the project further, Marker moved to Mexico and was able to isolate 3 kg (6.5 lb) of progesterone from 9 tonnes (10 tons) of cabeza de negro collected from the wild by indigenous Mexicans. Another species of yam, barbasco (*Dioscorea composita*), was subsequently also used as it contains five times as much diosgenin. With an inexpensive source of diosgenin available to make progesterone, chemists started to modify its structure in order to make other hormones such as testosterone, estradiol and estrone. In 1951, norethisterone (norethindrone), the first oral contraceptive for women, was synthesized. This can mimic the action of progesterone to stop ovulation during pregnancy, and is better absorbed from the digestive tract.

A method for converting diosgenin to anti-inflammatory corticosteroid drugs, for use in the treatment of inflammatory conditions such as rheumatoid arthritis, was also developed around this time. Today, corticosteroids are commonly used externally as anti-inflammatories for conditions such as dermatitis (see page 162).

As the demand for diosgenin increased, many thousands of Mexicans used their expert knowledge to collect yam roots from the wild or cultivate them. In the 1990s, however, demand for Mexican yams all but ended as synthetic methods to produce diosgenin were developed, and cheaper plant sources, such as the seeds of fenugreek (*Trigonella foenum-graecum*) in the legume family, became available.

The genus name *Dioscorea* honours the Greek physician Pedanius Dioscorides (*c.* 40–90 CE), whose *De Materia Medica* was probably the first European book of medicinal plants (see page 14). There are more than 600 species of yam distributed throughout the tropical and temperate regions of the world, a number of which are under threat and at risk of becoming endangered in the wild. While many

of them have edible tubers, others – including those used for their diosgenin content – are very bitter. More than a hundred species are known to be used medicinally.

LEFT **Pedanius Dioscorides.**

BELOW **The seeds of fenugreek (*Trigonella foenum-graecum*) are used in systems of traditional medicine, including Ayurveda, and are also a culinary spice.**

COTTON CONTRACEPTIVE

PLANT:

Gossypium hirsutum L.

COMMON NAME(S):

upland cotton, Mexican cotton

FAMILY:

mallow (Malvaceae)

ACTIVE COMPOUND(S):

NATURALLY OCCURRING:

bis-sesquiterpene (gossypol)

MEDICINAL USES:

MAIN: male antifertility

OTHER: to regulate menstruation

PARTS USED:

seed oil

Gossypol

ABOVE **Cotton (*Gossypium hirsutum*) seeds contain gossypol, which has been studied as a potential male contraceptive. It also shows other effects, including anti-protozoal and antiviral activities.**

In the first half of the twentieth century, the birth rate in a community in Jiangsu province in China dropped to zero. It was a cotton-producing area that over that period had changed from cooking with soybean oil to using crude cottonseed oil. On investigating the infertility, scientists found that the local men had a very low sperm count and women had disrupted menstruation. Could they have found a contraceptive for men?

Several species of cotton are cultivated worldwide, but around 80 per cent are cultivars and hybrids of upland cotton (*Gossypium hirsutum*). In China, where cotton is not native, other cultivated species include Levant cotton (*G. herbaceum*), which probably originated in Africa or western Asia. Cottonseed oil consists mainly of fatty acids (see box), but in addition it contains up to 1.3 per cent gossypol, a dimeric sesquiterpene that was identified in 1886 by chemists looking for potential dye compounds. The Chinese studies showed that the antifertility effects were caused by gossypol, and a series of trials involving thousands of volunteers in 18 provinces across the country followed.

SAFETY CONCERNS

The trials found that although a daily dose of gossypol resulted in reliable male contraception, there were some side effects of long-term use. The time it took for

fertility to return once gossypol treatment had ceased varied from several months to a few years, and was associated with the duration of the treatment. In up to 10 per cent of men, fertility did not return at all. Long-term use of gossypol was also linked to low potassium levels (hypokalaemia), with associated symptoms of muscle weakness, severe fatigue and paralysis. The relationship between gossypol and hypokalaemia was not straightforward, however, as the number of people who experienced it varied between trials from less than 1 per cent to almost 9 per cent.

Due to the observed side effects, further research into gossypol was largely discontinued in 1986, but interest did not disappear completely. Cottonseed oil contains both (-)-gossypol and (+)-gossypol, and it is the former that has the antifertility action. It affects how sperm mature and their ability to move, resulting in a zero sperm count. Modifications to the structure of gossypol have been created and tested, but so far none has had better antifertility effects than gossypol. The purity, form, dosage and duration of treatment continue to be studied to see if a safe and effective protocol can be found. There is still hope for a male 'pill' from cotton.

ABOVE Levant cotton (*Gossypium herbaceum*) is a perennial shrub that can reach 1.8 m (6 ft) tall. It is cultivated, including in China, but makes up less than 10 per cent of commercial production.

Soft fibres

Cotton is best known for the fibre made from its fine seed hairs, which gives us cotton fabric and cotton wool. In fact, the word cottons was first applied to soft fabrics made from low-grade sheep's wool. Only later was the term cotton wool used for fibres from a plant that came to be known as the cotton plant. The genus name *Gossypium* is derived from the Arabic word *goz*, meaning 'soft'.

The seeds themselves contain cottonseed oil, which principally consists of linoleic, palmitic and oleic fatty acids, and is used as a solvent for injections, in soaps and as a cooking oil. The seedcake and whole seeds are a protein-rich food for humans and livestock. However, cases of poisoning (particularly in pigs) have resulted from eating the seeds, due to the presence of gossypol. In contrast to the antifertility properties of cottonseed oil, the toxicity is due mainly to (+)-gossypol, which is usually present in larger amounts than (-)-gossypol. Gossypol can be removed by processing and cultivars that lack the compound have also been bred, although these are more susceptible to attack from pests.

ABOVE Seeds of cotton (*Gossypium hirsutum*) are the source of an oil that can disrupt male and female fertility.

LEFT Cotton (*Gossypium hirsutum*) fruit split open when ripe to expose the fine seed hairs that are the source of the cotton fibre.

EUROPE

Europe has land borders with Asia and the Middle East, and sea borders with the Arctic and Atlantic Oceans and the Mediterranean Sea. Consequently, its climate is broadly divided into continental, oceanic and Mediterranean types. Before the arrival of humans, the region was largely covered by forest, most of which has since been cleared. Conventional medicine is now used by most of the population, but herbal medicine persists and is seeing a resurgence in interest.

ABOVE **Monasteries and convents in Europe cultivated herbs to provide medicines for their infirmary, as well as to flavour food, and to freshen and clean interiors.**

The Middle Ages and monasteries

In 77 CE, Pedanius Dioscorides (c. 40–90 CE; see page 14), a Greek botanist and physician in the Roman army completed his work *De Materia Medica*, which outlined 600 plants used medicinally in the eastern Mediterranean. It was translated into several European languages and influenced herbal medicine in the region for more than 1,500 years. In fact, Mediterranean herbs are still commonly used in European cooking and herbal medicine (see box).

In the Middle Ages, monasteries in Christian regions of Europe acted as hospitals. Their medical knowledge was derived from local folk medicine, practical experience and copies made by monks of Arabic translations of Graeco-Roman medical texts (see page 85). Most of the plants they needed were grown in the monastery herb garden.

We know of several medical works written by monks, as well as *Physica*, the first herbal written by a woman, the German Benedictine nun and abbess Hildegarde of Bingen (1098–1179). These often survive only as later, modified copies. In southern Europe, in particular, the works of Arab scholars such as Ibn Sina (Avicenna) (c. 980–1037; see page 85) were also influential. Other regional differences include the *iatrosophia* (from the Greek for 'wisdom of healing'), medical texts produced in the Byzantine Empire (Eastern Roman Empire), probably from as early as the tenth century. They continued to be used by Greek Orthodox monasteries during the Ottoman Empire following the conquest of Constantinople (Istanbul) in 1453, right up until the early twentieth century.

The printing press

From the thirteenth century, wealthier members of European society increasingly used exotic spices for medicinal as well as culinary purposes, while poorer people still had to rely on herbs. The demand for spices fuelled the voyages of exploration and colonial conquest at the end of the Middle Ages.

Another impetus for change came from the development of the printing press in western Europe in the mid-fifteenth century. This made it possible for medicinal information to be available more cheaply and in the local language. Hildegarde of Bingen's *Physica*, for example, was published in book form in 1533. The work of another German, Leonhart (Leonhard) Fuchs (1501–1566; after whom the genus *Fuchsia* is named), challenged the prominence of Dioscorides. Published in 1542, his *De Historia Stirpium* described 400 plants from northern Europe. Later important herbals were written by English herbalists John Gerard (*c.* 1545–1612) and Nicholas Culpeper (1616–1654), who are quoted elsewhere through this book.

In Europe today, traditional medicine includes an estimated 1,300 plants from the region as well as another 700 from around the world. Many are collected from the wild in eastern Europe, but consumer demand has threatened the survival of some species. Conservation initiatives and increasing cultivation of medicinal plants is helping to preserve the biodiversity of the region.

LEFT **In Vision 4, 'Cosmos, Body, and Soul: The Word Made Flesh', of her medieval manuscript *Liber Divinorum Operum* (1163–1173), Hildegard of Bingen (seen bottom left) depicted the seasons of the year.**

MEDITERRANEAN INFLUENCE

Many plants originating in the Mediterranean would have been found in medieval gardens throughout Europe and are still widely used today in herbal medicine and sometimes also cooking. Members of the daisy family (Asteraceae) include pot marigold (*Calendula officinalis*), whose bright orange 'petals' (ray florets) can be added to salads. They are used externally in creams and oils for minor wounds, insect bites and inflammation. Feverfew (*Tanacetum parthenium*) is also in the daisy family and is native to Greece and eastwards to Kashmir. Traditional uses for the leaves include being taken to aid digestion, and for pain and fever. The anti-inflammatory and pain-relieving properties of feverfew have been studied, and the plant is of particular interest to help prevent migraine attacks and for rheumatoid arthritis. Parsley (*Petroselinum crispum*; see page 29) in the carrot family (Apiaceae) is another Mediterranean plant. Traditional uses for the leaves and roots included as a diuretic, for gout and for kidney stones.

ABOVE **Flowering feverfew (*Tanacetum parthenium*) plants.**

Premenstrual syndrome

After puberty and before the onset of the menopause (see page 194), a woman's body has a monthly cycle, during which the levels of a number of hormones fluctuate. A collection of symptoms known as premenstrual syndrome (PMS) can result from an imbalance in these hormones. Herbal preparations such as chasteberry, and food supplements containing *gamma*-linolenic acid, may offer some relief.

CHASTEBERRY

PLANT:
Vitex agnus-castus L.
COMMON NAME(S):
chasteberry, chaste tree,
agnus-castus, monk's pepper
FAMILY:
mint (Lamiaceae)

ACTIVE COMPOUND(S):
NATURALLY OCCURRING:
diterpenes (rotundifuran);
flavonoids (casticin)
MEDICINAL USES:
MAIN: premenstrual
syndrome
OTHER: menstrual
irregularities, menopause
PARTS USED:
fruit, aerial parts

Records of chasteberry being used medicinally and symbolically date back at least 2,500 years (see box). However, the modern applications of this Mediterranean shrub in Europe and, more recently, North America, are associated almost exclusively with herbal preparations for the relief of PMS and other conditions related to the menstrual cycle. There is some evidence to support the use of chasteberry for PMS, although it is not clear which compounds are responsible for any beneficial effects and more than one are likely to be involved.

Laboratory studies found that an extract of the plant and its diterpene constituents could bind to dopamine receptors, with the diterpene rotundifuran consequently being able to inhibit the secretion of the hormone prolactin. Clinical studies of chasteberry in women with abnormally high levels of prolactin

Monk's pepper

Chasteberry has long been associated with chastity, hence its common name. Chasteberry flowers were worn by women who remained 'pure' during Greek festivals to honour Demeter, the goddess of fertility, and in ancient Rome the Vestal Virgins carried them for the same reason. Later, Christian monks walked over strewn chasteberry flowers during their initiation ceremony. Linked to this is one of the chasteberry's historical applications, as recommended by the Greek physician Dioscorides, which was to reduce sexual desire (libido). Monks added the ground spicy fruit to their food for this reason, giving rise to the plant's alternative common name of monk's pepper.

LEFT Chasteberry (*Vitex agnus-castus*) fruit resemble peppercorns, which gave rise to the plant's alternative common name of monk's pepper.

(hyperprolactinaemia) have provided some evidence to support a reduction in prolactin levels. As slightly raised levels of prolactin have been linked to PMS, as well as irregular menstruation and tender breasts, reducing prolactin levels may be one mechanism by which chasteberry could exert its effects. While clinical studies have, on the whole, indicated that chasteberry can alleviate symptoms of PMS, more studies are needed to confirm this, as well as the mechanisms and active compounds involved.

Another traditional use for chasteberry was to promote the production of breast milk (lactation). There are no laboratory or clinical studies to support this, however, and it seems likely that, due to chasteberry's action of lowering levels of prolactin, it would have the opposite effect and inhibit lactation.

EVENING PRIMROSE OIL

Gamma-linolenic acid (GLA) is an omega-6 unsaturated fatty acid and a precursor to compounds called prostaglandins, which have many roles in the body. GLA may be present in the diet in small amounts, but the human body can also produce it from linoleic acid, which is usually found in sufficient quantities in cooking oils. In some circumstances, however, such as when conversion from linoleic acid is inefficient, there may be a deficiency in GLA. The seeds of some plants, including evening primrose, produce an oil that is a rich source of GLA. Evening primrose oil has been of interest for the relief of PMS symptoms, tender breasts, menopause and many other conditions, and is now one of the most popular food supplements worldwide. Studies on its use for PMS have had mixed results, however, with only some showing benefits.

LEFT Chasteberry (*Vitex agnus-castus*) shrubs have palmately divided leaves and carry long heads of mauve flowers that are attractive to bees and butterflies.

RIGHT Evening primrose (*Oenothera biennis*) is a North American plant whose main medicinal use by Native Americans was as a poultice for wounds and skin problems. All parts of the plant were also eaten as a vegetable.

Easing the menopause

At the end of the fertile period of a woman's life – usually when she is in her late 40s or early 50s – levels of oestrogen (estrogen) and progesterone decrease significantly, while those of other hormones increase. This is the menopause, and it can result in a range of symptoms that include hot flushes, night sweats and sleep disturbance. The pharmaceutical answer is hormone replacement therapy, but plant-based remedies are popular alternatives.

BLACK COHOSH

PLANT:
Actaea racemosa L. (syn. *Cimicifuga racemosa* (L.) Nutt.)

COMMON NAME(S):
black cohosh, black snakeroot, black bugbane

FAMILY:
buttercup (Ranunculaceae)

ACTIVE COMPOUND(S):

NATURALLY OCCURRING:
triterpene glycosides (actein, 27-deoxyacetin, several cimicifugosides); some studies report the occurrence of isoflavonoid (formononetin)

MEDICINAL USES:

MAIN: menopause

OTHER: premenstrual syndrome

PARTS USED:
root and rhizome

Black cohosh is a North American herbaceous perennial in the buttercup family that can reach 2 m (6 ft) in height when in flower. Native Americans had many medicinal uses for the plant (see box), and it is for one of these – the relief of menopausal symptoms – that it is now a widely used herbal remedy.

There have been many studies into the effectiveness of black cohosh herbal remedies, some of which have shown evidence that it provides benefit for hot flushes, sweating, sleep disorders and nervous irritability. As with all herbal preparations, there is probably more than one compound that contributes to these effects, which in the case of black cohosh include a number of triterpene glycosides. The mechanism is not fully understood, but it is at least partly due to an ability to

LEFT **Although other species of *Actaea* have been used to repel insects, giving rise to both the common name of bugbane for the genus and the scientific name *Cimicifuga* (in Latin, *cimex* means 'insect' and *fugare* means 'to drive away'), the flowers of black cohosh (*A. racemosa*) are pollinated by insects, including bumblebees.**

reduce the concentration of luteinizing hormone circulating in the body – this is one of the hormones that normally increases during the menopause and is thought to be responsible for some of the associated symptoms. There have, however, been concerns over the safety of products listing black cohosh among their ingredients due to cases of serious adverse effects following their use. At least some of these products were found to contain species of *Actaea* from East Asia (see page 63) that contain different active compounds.

Black cohosh has a fairly restricted natural range, growing in eastern North America in moist woodland soil, usually in deep shade. The plant is now grown commercially in Europe, but most of the material used in herbal products is still harvested from the wild population. Considerable quantities are collected, with one estimate putting the figure at more than 1,000 tonnes (1,100 tons) of root and rhizome between 1995 and 2007. Combined with habitat loss, this has led to the decline of the species in the wild, causing concern for its survival in several American states.

PHYTOESTROGENS MIMIC OESTROGEN

Certain plant compounds – including some isoflavonoids, coumestans and lignans – can bind to the same receptors as the hormone oestrogen in humans and are therefore known as phytoestrogens. Members of the legume family are particularly rich in

ABOVE **Red clover (*Trifolium pratense*) is widely grown as a 'green manure'. Like many other members of the legume family, it fixes atmospheric nitrogen via bacteria in root nodules.**

isoflavonoids, and diets that include foods made from soybeans (*Glycine max*), which are naturally high in phytoestrogens, are said to offer some protection against symptoms of the menopause. Red clover (*Trifolium pratense*) is another source of isoflavonoids, so is widely used as an ingredient in food supplements aimed at women during the menopause. It is also of interest for benign prostatic hyperplasia in men (see page 110).

Black and blue

Cohosh is a word from the Native American Algonquian languages meaning 'rough', and refers to the rough roots of blue cohosh (*Caulophyllum thalictroides*), another medicinal plant in the barberry family (black cohosh has smooth roots).

Native Americans used black cohosh roots for numerous medical conditions, including kidney problems, tuberculosis, sore throat, rheumatism, fatigue and snakebites, the last leading to the plant's alternative common name of black snakeroot. The Europeans who colonized North America adopted many of these remedies, particularly those used for menstrual and menopause symptoms. Although the popularity of black cohosh declined for a while, it is now widely used in North America, Europe and elsewhere.

LEFT **Black cohosh (*Actaea racemosa*) was an ingredient of patent medicines such as Lydia E. Pinkham's Vegetable Compound. Bethroot or birthroot (*Trillium erectum*), in the trillium family (Melanthiaceae), was another and was also one of the plants from which diosgenin was later isolated (see page 187).**

Enhancing performance

Impotence is a common problem affecting many men at some point in their life. A number of pharmaceutical drugs have been developed to treat impotence, but there are also plants that are marketed as 'herbal Viagra'. Most focus has been on yohimbe and one of its alkaloids, yohimbine, but other plants are also of interest, including horny goat weed.

AFRICAN APHRODISIAC

PLANT:

Corynanthe johimbe K.Schum. (syn. *Pausinystalia johimbe* (K.Schum.) Pierre)

COMMON NAME(S):

yohimbe

FAMILY:

coffee (Rubiaceae)

ACTIVE COMPOUND(S):

NATURALLY OCCURRING:

indole alkaloids (yohimbanes: yohimbine [corynine, quebrachine])

MEDICINAL USES:

MAIN: impotence

OTHER: aphrodisiac

PARTS USED:

bark

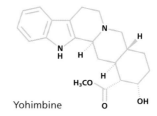

Yohimbine

ABOVE **Yohimbe (*Corynanthe johimbe*) bark contains indole alkaloids, including yohimbine. Extracts from its bark have been used traditionally for their reputed aphrodisiac effects.**

ABOVE **Yohimbe (*Corynanthe johimbe*) bark.**

Yohimbe is a large tree from the coastal forests of central West Africa, whose bark is traditionally used as a male tonic and aphrodisiac. For the last century, it has been available as a food supplement in Europe, and it is now one of the top-selling herbal supplements in North America. Interest in this plant is not restricted to erectile dysfunction associated with impotence, and claims are also made that it aids sports enhancement and weight loss. In recent years, however, many countries have imposed restrictions or complete bans on the sale of products containing yohimbe owing to safety concerns.

The active compounds in yohimbe are indole alkaloids, with yohimbine being present in the greatest, but variable, amounts. Some studies have been carried out on this compound and it has been developed as a pharmaceutical drug in the United States, although it is now rarely prescribed. Yohimbine blocks *alpha*2-adrenoceptors and, to a lesser extent, *alpha*1-adrenoceptors, principally affecting peripheral nerve signals in the body but also with some effects on the central nervous system. This can result in dilation of the peripheral blood vessels (vasodilation), which is thought to be the main mechanism for any action on male impotence. Although such effects are usually modest, there have also occasionally been reports of priapism (a painful, enduring erection). Even at low doses, yohimbine can cause symptoms such as an increase in blood pressure and heart rate, sweating, anxiety, agitation and tremors.

HORNY GOAT WEED

The leaves of several species of barrenwort (*Epimedium* spp.) are sold as supplements under the eye-catching name of horny goat weed, which originates from a translation of their name in traditional Chinese medicine, *yin yang huo*. According to legend, a goat herder noticed that his animals became more sexually active after eating these particular plants. Today, *yin yang huo* is widely prescribed in traditional Chinese medicine for impotence, as an aphrodisiac for both men and women, and for PMS, the menopause and osteoporosis. The leaves of the plants contain flavonol glycosides such as icariin. Laboratory studies suggest that icariin has a similar mechanism of action to sildenafil, a pharmaceutical drug used for impotence, producing an inhibitory effect on the enzyme phosphodiesterase type 5. However, more studies – including clinical trials – are required to evaluate the usefulness of horny goat weed in impotence and other conditions.

BELOW *Yin yang huo*, a drug used in traditional Chinese medicine, can be sourced from the leaves of four different species of barrenwort (*Epimedium* spp.), namely *E. brevicornu*, *E. koreanum*, *E. pubescens* and *E. sagittatum* (pictured). It can be processed by stir-frying the leaves in mutton fat, which enhances the drug's effectiveness.

Supply and demand

In addition to the local use of yohimbe bark, there is high international demand for extracts due to the claims it can enhance sexual and sporting performance. The yohimbe tree is found in forests from southeast Nigeria to Gabon, but the majority of material is collected from Cameroon. It provides an additional income for local people, who often harvest it in conjunction with logging for other timbers. Despite yohimbine being present in other parts of the plant, only the bark is usually used and the whole tree is often felled in the process. The reasoning behind this is that taking strips of the bark enables woodboring insects to attack the tree, so it would die anyway. There have been concerns over the sustainability of this supply for many years, particularly as the trees are not very common. Despite this, trade in yohimbe is not currently controlled by the Convention on International Trade in Endangered Species of Wild Fauna and Flora.

CHAPTER 10

FIGHTING CANCER

Cancer is a group of diseases in which the cells in a certain part of the body multiply in an uncontrolled way. These cancer cells can then invade or spread to healthy parts of the body. Therapeutic strategies include the use of chemotherapy drugs to inhibit the proliferation of cancer cells or destroy them. Plants have had a key role in drug discovery for cancer. This chapter describes some important chemotherapy drugs that are derived from natural origins.

Madagascar periwinkle (*Catharanthus roseus*)

NATURAL COMPOUNDS FOR FIGHTING CANCER

Worldwide, cancer is the second leading cause of death after cardiovascular disease, and according to the World Health Organization, 9.6 million people died from cancer in 2018. The causes of cancer are often unknown and can be complex, and there are many different types. Current chemotherapy drugs aimed at treating cancer include those derived from plants, while there is continued interest in how plants in our diet might offer protection against some forms of cancer.

PLANTS AS POWERHOUSES

Certain plants produce toxic chemicals to deter predators and protect themselves from being eaten. Many – although not all – of these chemicals are alkaloids. Drugs used in chemotherapy regimens to treat cancer often inhibit the mechanisms that enable cancer cells to proliferate. Because certain plants produce powerful chemicals with toxic effects, humans have harnessed these from nature to develop drugs that destroy cancer cells.

PLANT-DERIVED CHEMOTHERAPIES

Records of the use of plants to combat cancer date back to the Ebers Papyrus in 1500 BCE. Since then, plants have continued to be used for their reputed anticancer effects – for example, the anti-tumour effects of the mayapple (*Podophyllum peltatum*; see page 202) in the barberry family (Berberidaceae) were documented in the 1860s. Another plant in this genus, the Himalayan mayapple (*P. hexandrum*; see page 202), has become the preferred source of podophyllotoxin, which is used for

the semi-synthesis of anticancer drugs such as etoposide. Other chemotherapy drugs derived from plants include paclitaxel, which was originally discovered in the bark of the Pacific yew (*Taxus brevifolia*; see page 206) in the yew family (Taxaceae), and vincristine from the Madagascar periwinkle (*Catharanthus roseus*; see page 210) in the dogbane family (Apocynaceae). The happy tree (*Camptotheca acuminata*; see page 208) in the tupelo family (Nyssaceae), used in traditional Chinese medicine, has provided the alkaloid camptothecin, which was used to develop chemotherapy drugs such as topotecan.

TARGETING THE SKIN

The potent biological activities of some plant chemicals have been developed specifically for the condition actinic keratosis, which is thought to be the start of some skin cancers. Drugs developed for topical use for this condition include a complex chemical from the sap of petty spurge (*Euphorbia peplus*; see page 214), a member of the spurge family (Euphorbiaceae). The names of this genus and family mean 'good pasture' and are derived from Euphorbus, a physician to King Juba II of Numidia (50 BCE–23 CE), now known as Algeria. Other topical medicines developed for actinic keratosis include masoprocol from the creosote bush (*Larrea tridentata*; see page 214) in the caltrop family (Zygophyllaceae), and solasodine glycosides from members of the potato family (Solanaceae). One of the solasodine glycosides is solamargine, which can be found in the purple African nightshade (*Solanum marginatum*), a plant native to tropical northeast Africa that has been used for a range of traditional medicinal purposes (see page 215).

POSSIBILITIES FOR PREVENTION

The World Health Organization reports that a number of lifestyle and dietary factors, including a low intake of fruits and vegetables, may contribute to the risk of developing some cancers. In line with this, there has been considerable interest in the role plants might play in our diet to prevent cancer. Many dietary plants have been studied in this respect, including turmeric (*Curcuma longa*; see page 216) in the ginger family (Zingiberaceae), and cruciferous vegetables such as cabbage and broccoli (*Brassica oleracea*; see page 217) in the cabbage family (Brassicaceae).

LEFT Polarized light micrograph of paclitaxel crystals. This chemotherapeutic drug is derived from the bark of Pacific yew (*Taxus brevifolia*).

ABOVE Maytansine, a compound used in breast cancer therapy, is present in some members of the spindle family (Celastraceae; see page 212), including false spike-thorn (*Putterlickia pyracantha*), a shrub from the Cape Provinces of South Africa.

ABOVE Purple African nightshade (*Solanum marginatum*), a source of solamargine, has defensive thorns on its leaves and stems.

Plant chemicals and chemotherapy

Plants have played a key role in the discovery of cancer chemotherapy drugs, providing unique and important chemicals. These act in different ways to kill or inhibit cancer cells. They include vincristine from the Madagascar periwinkle, and paclitaxel, originally from the Pacific yew. Other examples of drugs developed from plant chemicals include etoposide, which is semi-synthesized from podophyllotoxin from the Himalayan mayapple, and topotecan, derived from camptothecin that occurs in the happy tree.

THE HIMALAYAN MAYAPPLE

PLANT:
Podophyllum hexandrum Royle (syn. *Sinopodophyllum hexandrum* (Royle) T. S. Ying)
COMMON NAME(S):
Himalayan mayapple, Indian mayapple
FAMILY:
barberry (Berberidaceae)

ACTIVE COMPOUND(S):
NATURALLY OCCURRING:
lignans (podophyllotoxin)
SEMI-SYNTHESIZED:
etoposide, teniposide
MEDICINAL USES:
MAIN: cancer chemotherapy
OTHER: warts
PARTS USED:
rhizomes and roots

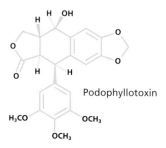

Podophyllotoxin

LEFT The resin from Himalayan mayapple (*Podophyllum hexandrum*) rhizomes contains lignans, including podophyllotoxin, which is used for the semi-synthesis of anticancer drugs such as etoposide and teniposide.

BELOW Himalayan or Indian mayapple (*Podophyllum hexandrum*) flowering plants. Flowers appear singly at the top of short stems before the leaves are fully developed.

The story of the anticancer drugs etoposide and teniposide began with studies on the North American mayapple. In 1861, the anti-tumour properties of mayapple rhizomes were reported by botanist Robert Bentley (1821–1893) at King's College London. Twenty years later, the lignan podophyllotoxin was isolated from the rhizomes. But it was not until the 1940s that this compound was discovered to prevent cancer cell division.

Podophyllotoxin was too toxic to be developed as a medicine for internal use. Instead, its chemical structure was used to design new compounds in the hope of finding other anticancer drugs. Two of these new compounds, etoposide and teniposide, were produced by partial synthesis from podophyllotoxin itself. Interestingly, these drugs had a new mode of action – they bind to an enzyme called topoisomerase II, which is required when DNA unwinds during cell division. By preventing the enzyme from doing this job, etoposide and teniposide interfere with DNA synthesis in cancer cells, so are toxic to them. Etoposide is now used in cancer chemotherapy for certain cancers, including some forms of lung cancer and lymphomas. Teniposide is used as an anticancer drug for some types of lymphoma and for other cancers.

To source podophyllotoxin as a topical medicine for warts (see page 172) and for the production of these anticancer drugs, a new solution was found to meet the demands of the pharmaceutical industry. The Himalayan mayapple, which has a native range extending from east Afghanistan to China, contains higher concentrations of podophyllotoxin (4 per cent)

in the rhizome, compared to the mayapple (0.25 per cent). Consequently, the Himalayan mayapple became the preferred source of podophyllotoxin for drug development. But the story of its role in medicines history does not end there.

PROTECTING OUR PLANET

Early research into medicines from plants did not often consider the consequences of overharvesting plants from the wild and exploiting biodiversity. Today, the field has evolved so that plant conservation strategies are now in place, including international agreements between governments and legislation to protect the biodiversity of our planet.

Podophyllotoxin cannot be synthesized efficiently from scratch in the laboratory, so is sourced from the Himalayan mayapple for drug manufacture. This puts pressure on natural populations of the plant, which have been drastically reduced, and the species is becoming at risk of extinction in the wild. The trade of the plant is therefore now restricted under the Convention on International Trade in Endangered Species of Wild Flora and Fauna. Other solutions to conserve wild populations of the Himalayan mayapple are also under evaluation (see box).

Witches' umbrella

Because wild populations of the Himalayan mayapple are under threat, alternative sources of podophyllotoxin for pharmaceutical drug development are being sought. One approach being studied is the development of plant cell cultures to produce podophyllotoxin. There has also been renewed interest in obtaining this important plant compound from the American species of mayapple, which according to folklore was once used as a poison by witches. This explains why the plant has been known as witches' umbrella and devil's apple. There has also been particular interest in sourcing podophyllotoxin from the leaves of mayapple rather than its rhizomes, as their harvest need not kill the plant.

ABOVE Mayapple (*Podophyllum peltatum*) growing wild in North America. Its umbrella-shaped leaves arise from creeping rhizomes.

NORTH AMERICA

The region of North America north of Mexico has a flora of around 22,000 plants, growing in habitats ranging from cold and warm deserts, to subarctic evergreen forest and temperate rainforest, and to mountain steppes and prairies. The native people of North America found medicinal uses for at least 4,000 species, and some of these plants now provide herbal preparations and pharmaceutical drugs with a much wider reach.

RIGHT **An advertisement for Shaker Family Pills, 1891, claiming their effectiveness in curing sick headaches, constipation, a sluggish liver and the effects of biliousness.**

The eclectics

Native North Americans relied on local plants to provide medicines, but because this knowledge was an oral tradition, passed directly from one person to another, there is no surviving evidence of how it developed. European settlers arriving on the continent from the sixteenth century onwards brought their own medicines with them, but over time they incorporated many local plants and their traditional uses after observing the practices of the indigenous people. A member of the Shaker religious sect asked, 'Why look to Europe's shores for plants that grow at our own doors?', and a group at Sabbathday Lake in Maine set up a herb business in 1799.

In the early nineteenth century, some doctors abandoned more drastic European practices such as the use of bleeding, mercury and arsenic in favour of North American herbs combined with the local practices of sweating, purging and emetics. This was also the era of patent medicines. In the 1820s, a group of doctors practised what came to be known as eclectic medicine. They introduced many indigenous plants that are now well known, and several active resins. One of the first resins to be introduced, podophyllin from mayapple (*Podophyllum peltatum*) in the barberry family, is still in use today (see page 172). By 1895, there were 10,000 physicians practising eclectic medicine in North America, but just 40 years later it had been largely superseded by European herbal medicine. In its turn, European herbal medicine – partly driven by German interest and research – has incorporated many North American herbs. Today, more than 180 North American plants are harvested from the wild for use around the world.

Eastern forests and southwestern deserts

The deciduous forests of the Appalachian and Ozark mountains provide a wealth of medicinal plants that are now familiar around the world. The rhizomes of goldenseal (*Hydrastis canadensis*) in the buttercup family (Ranunculaceae), for example, are used traditionally for digestive complaints. Witch hazel (*Hamamelis virginiana*),

ABOVE **Flowering goldenseal (*Hydrastis canadensis*) plants.**

in the witch hazel family (Hamamelidaceae), is just one of the plants that once made up an ancient northern hemisphere flora (see page 62). Its bark (which contains astringent tannins) and leaves were used externally by Native Americans to alleviate aching muscles, inflammation and sores, and internally for diarrhoea and coughs.

A second important area for North American medicinal plants is the arid and semi-arid deserts in the southwest of the continent. Creosote bush or chaparral (*Larrea tridentata*; see pages 97 and 215) in the caltrop family and Mormon tea (*Ephedra nevadensis*; see page 118) in the ephedra family (Ephedraceae) are among the species found there. Ephedrine and related alkaloids are found in other species of the genus *Ephedra* (see page 120), but their presence in Mormon tea is in question. Native Americans made a tea from stems of this plant and introduced European settlers to the practice, and the species consequently became known by many common English names, including desert tea and teamster's tea.

PRAIRIE PLANTING

One of the most well known North American medicinal plants is the purple coneflower (*Echinacea purpurea*; see page 126) in the daisy family (Asteraceae), and together with two other species in the genus it is known as the herbal remedy echinacea. Purple coneflower is native to the prairies of middle America, and it shares this habitat with a number of other plants that are both medicinal and herbaceous stalwarts of temperate gardens. Culver's root (*Veronicastrum virginicum*, syn. *Leptandra virginica*) in the speedwell family (Plantaginaceae) was the source of the eclectic resin leptandrin, used as a purgative. Wild bergamot (*Monarda fistulosa*) in the mint family (Lamiaceae) contains geraniol and thymol; the latter is diluted to give an antiseptic mouthwash.

ABOVE **Culver's root (*Veronicastrum virginicum*) was used traditionally by Native Americans, including as a purgative.**

LEFT **Flowering witch hazel (*Hamamelis virginiana*) shrub – its common name comes from wych, an old English word meaning 'tree with bendable branches'. Extracts of the twigs, bark and leaves are now widely used externally for inflammation and irritation.**

THE WONDER OF YEW

PLANT:
Taxus brevifolia Nutt.

COMMON NAME(S):
Pacific yew

FAMILY:
yew (Taxaceae)

ACTIVE COMPOUND(S):

NATURALLY OCCURRING:
diterpene (paclitaxel)

SEMI-SYNTHESIZED:
docetaxel, cabazitaxel

MEDICINAL USES:

MAIN: cancer chemotherapy

OTHER: poison

PARTS USED:
bark

Paclitaxel

ABOVE **The bark from the Pacific yew (*Taxus brevifolia*) was the original source of the anticancer drug paclitaxel, but this and related drugs are now more sustainably produced by semi-synthesis using precursor chemicals from the leaves of the European yew (*Taxus baccata*).**

Throughout history, yew trees (*Taxus* spp.) have been shrouded in myth and magic, and their poisonous properties have been feared. Indeed, the Roman scholar Pliny the Elder (23–79 CE) declared that those who drank wine from casks made of yew wood died. Another ancient belief was that just sleeping in the shadow of a yew tree would result in sickness or even death. There are also historical reports of beekeepers putting their hives near yew trees and ending up with poisonous honey. Although the toxic effects of yew trees have long been known, these potent properties can give us clues to discover toxic chemicals to target cancer cells for therapeutic use.

The Pacific yew is native to south Alaska and the west of the United States. In the 1960s, bark extracts were first studied for their anticancer effects. After promising laboratory results, a new chemical with novel anti-tumour and anti-leukaemia effects was isolated in 1971. Initially called taxol and later renamed paclitaxel, it was discovered to have a unique mode of action. Paclitaxel stabilizes microtubule assembly in cells, so essentially blocks the cycle of cancer cells, inhibiting their replication.

Although paclitaxel was a promising drug candidate, yields from the bark were low and thousands of trees were needed to obtain enough of the drug for pharmaceutical use, putting the survival of the species at risk. Indeed, harvesting the bark from the Pacific yew contributed to serious consequences on

tree populations, which have declined by approximately 30 per cent within the last three generations. The Pacific yew is now classified on the International Union for Conservation of Nature Red List as Near Threatened. The chemical structure of paclitaxel was too complicated to be made from scratch in the laboratory, so a more sustainable way of obtaining this useful anticancer drug without overharvesting trees had to be found.

SAVING BIODIVERSITY

Knowledge of plant systematics can tell us how plants are related to each other. If we know that one plant contains useful chemicals, studying closely related species may help us predict which of these could contain other important chemicals, such as those for medicinal applications. Scientific research revealed that the leaves and twigs of the European or English yew (*Taxus baccata*) contain precursor chemicals, such as 10-deacetylbaccatin III, which are similar to paclitaxel. These chemical precursors can be sustainably harvested from the leaves and twigs of cultivated trees, then converted by semi-synthesis in laboratory conditions to paclitaxel. This solution enabled adequate amounts of paclitaxel to be sourced, and it was approved as a cancer chemotherapy drug from 1992, including for use in ovarian and breast cancers. The chemical precursors from the European yew were not only useful to provide paclitaxel, but also in the development of other, new anticancer drugs. These include docetaxel, which is used for certain types of breast and lung cancer, and cabazitaxel, which has shown activity against tumours resistant to other taxane drugs such as paclitaxel.

LEFT **Close-up of the bark of a Pacific yew (*Taxus brevifolia*), a small evergreen tree that can reach 6–9 m (20–30 ft) in height and grows along the northwestern coast of North America.**

RIGHT **European yew (*Taxus baccata*) needles (leaves) contain a compound that can be converted to paclitaxel, and the regular trimming of hedges provides a ready source of material that can be used by the pharmaceutical industry. The fleshy red arils are non-toxic and are a source of food for birds and mammals.**

The long and winding road

In the same family as the Pacific and European yews is the Japanese plum yew (*Cephalotaxus harringtonia*; see page 22), which is native to Assam and Japan. In the 1970s, the root, stems and bark of the tree were found to contain alkaloid chemicals, including homoharringtonine, which were revealed to have potent anti-leukaemia activity. A semi-synthetic derivative of homoharringtonine, called omacetaxine, was also later discovered to inhibit protein synthesis, thereby promoting cancer cell death. However, it was more than 40 years after the first discovery of homoharringtonine's anticancer effects that omacetaxine was developed as an approved drug for certain forms of leukaemia. This is one example of how drug discovery can be a long and winding road.

THE HAPPY TREE

PLANT:
Camptotheca acuminata
Decne.

COMMON NAME(S):
happy tree

FAMILY:
tupelo (Nyssaceae)

ACTIVE COMPOUND(S):

NATURALLY OCCURRING:
quinoline alkaloids
(camptothecin)

SEMI-SYNTHESIZED:
topotecan, irinotecan

MEDICINAL USES:

MAIN: cancer chemotherapy

PARTS USED:
bark

Camptothecin

ABOVE **The anticancer quinoline alkaloid camptothecin was originally discovered in the bark of the happy tree (*Camptotheca acuminata*). It was used to inspire the development of new anticancer drugs such as irinotecan and topotecan.**

The happy tree has a native range stretching from Tibet to south China, and different parts of the tree have long been used in China as a traditional remedy for cancer. In the 1950s, the National Cancer Institute in the United States started a programme to discover new anticancer drugs, which included tests on plant extracts. The research discovered that Pacific yew was a source of useful anticancer drugs (see page 206), and another promising extract that proved toxic to cancer cells in laboratory studies was prepared from the bark of the happy tree. This catapulted the discovery of important drugs that remain valuable in cancer chemotherapy today.

In 1966, the main active constituent of happy tree bark was revealed to be the quinoline alkaloid camptothecin. Following this discovery, camptothecin was described as a novel leukaemia and tumour inhibitor, but although it showed potent anti-tumour activity when tested in people with cancer, it caused severe adverse effects. It was also poorly soluble in water, so was not ideal for development as a pharmaceutical drug. By the 1970s, it seemed that the happy tree was not going to become a source of anticancer drugs after all. However, almost 20 years after the discovery of camptothecin, its unique mode of action was unveiled, sparking renewed interest in this alkaloid as a template to design new anticancer drugs.

NO LONGER IN SHAPE

Camptothecin inhibits the enzyme topoisomerase I in cancer cells. Topoisomerase enzymes regulate DNA conformation (or shape), and facilitate processes such as DNA replication and repair. When topoisomerase I is inhibited by camptothecin, the enzyme cannot effectively perform these actions, so the DNA is damaged and not repaired. This results in the cancer cells being killed.

In the mid-1980s, more water-soluble analogues of camptothecin were synthesized in the hope of finding new and effective chemotherapy drugs. The first of those developed were irinotecan and topotecan, but they were not approved by the United States Food and Drug Administration until the mid-1990s. It took around 30 years from the discovery of camptothecin

LEFT **The happy tree (*Camptotheca acuminata*) may be so named because you feel happy when you see it. The discovery that the bark of this tall tree, which can reach 20–30 m (64–100 ft) in height, is the source of a useful cancer drug has depleted wild populations and it is now considered endangered.**

for these novel topoisomerase I-inhibiting drugs to become available. Yet this time and effort has ultimately improved the range of therapies available for people with certain cancers. Topotecan is used as chemotherapy for some forms of ovarian and cervical cancer, while irinotecan is used in treatment regimens for different types of colorectal and pancreatic cancers. Another drug based on camptothecin is belotecan, which is used for some types of ovarian and lung cancers. Other analogues of camptothecin are also being explored as potential anticancer drugs, including rubitecan, karenitecin, diflomotecan, gimatecan and exatecan.

ABOVE **Happy tree (*Camptotheca acuminata*) bearing spherical heads of elongated single-seeded fruit.**

Family similarities

Since the discovery of camptothecin in the happy tree, the alkaloid has been found to occur in a range of other plants in different families. These include members of the white pear family (Icacinaceae), namely ghanera (*Mappia nimmoniana*, syn. *Mappia foetida* and *Nothapodytes nimmoniana*), which has a native range extending from India eastwards to the Philippines and south to Sumatra; and *Merrilliodendron megacarpum*, which is native to the Philippines and west Pacific. Another source of camptothecin is *Ophiorrhiza pectinata* in the coffee family (Rubiaceae), which originates from central Sri Lanka. Nag kuda (*Tabernaemontana alternifolia*) in the dogbane family is native to western and southern India. It also contains camptothecin and has been used as a traditional remedy in India for snakebites. Another member of this family, the Madagascar periwinkle, also produces chemicals that have anticancer activity (see page 210).

ABOVE Ghanera (*Mappia nimmoniana*) is a tree that reaches 8–10 m (26–33 ft) in height and bears heads of fruit that ripen to a reddish purple. Its roots have been used medicinally in India.

THE MADAGASCAR PERIWINKLE

PLANT:

Catharanthus roseus (L.)
G.Don

COMMON NAME(S):

Madagascar periwinkle

FAMILY:

dogbane (Apocynaceae)

ACTIVE COMPOUND(S):

NATURALLY OCCURRING:

bisindole alkaloids
(vincristine, vinblastine)

SEMI-SYNTHESIZED:

vinorelbine, vindesine,
vinflunine

MEDICINAL USES:

MAIN: cancer chemotherapy

OTHER: diabetes

PARTS USED:

leaves

Vincristine

ABOVE **The Madagascar periwinkle (*Catharanthus roseus*) is a source of the anticancer drugs vincristine and vinblastine, and other alkaloids in the plant can be chemically modified to further increase yields of the drugs.**

BELOW **The native range of the Madagascar periwinkle (*Catharanthus roseus*) is restricted to the island after which it is named, but it has spread around the world and is widely cultivated for its attractive pink or white flowers.**

The study of a plant used traditionally as a remedy for diabetes in the West Indies led to the serendipitous discovery of important anticancer drugs that proved to be a medical breakthrough in the treatment of leukaemia. The plant is the Madagascar periwinkle, named because it is native to the Indian Ocean island. Extracts of its leaves were first tested in the 1950s, and were shown to decrease numbers of white blood cells (leucocytes) and, later, to have some effect against leukaemia in laboratory tests. Two potent alkaloids, vincristine and vinblastine, were subsequently isolated from the plant. In the 1960s, these compounds were tested in clinical trials in people with various forms of leukaemia and lymphoma, and were concluded to produce positive outcomes. They were also tested for their usefulness for various solid tumours, with more promising results.

Further scientific research revealed that vincristine and vinblastine act by binding to a protein in cells called tubulin. The complex formed between the alkaloid and tubulin inhibits the assembly of microtubules required for cell division. As a result, cancer cell division is prevented by these chemotherapy drugs. Today, they are used for various cancers, including leukaemia, lymphomas and some solid tumours, such as certain types of lung cancer. Both alkaloids are also used for the semi-synthesis of new anticancer drugs, including vinorelbine, vindesine and vinflunine. The first two of these are given to some patients with certain subtypes of breast and lung cancer, while vinflunine is used as a chemotherapy drug for some forms of bladder cancer.

PLANTS AS CLEVER CHEMISTS

The structures of vincristine and vinblastine are so complex that the chemicals cannot be synthesized in

the laboratory on a practical scale, so scientists need to rely on the plant to source these useful anticancer drugs. Vincristine and vinblastine are only minor constituents of the Madagascar periwinkle. For example, vincristine occurs at only 0.0002 per cent, so large quantities of plant material are needed to extract sufficient amounts for pharmaceutical use – it is estimated that 500 kg (1,100 lb) of plant material would be needed to produce just 1 g (0.04 oz) of vincristine. Furthermore, the process to extract and purify the alkaloids from the plant is very laborious and time consuming.

Because Madagascar periwinkle produces higher levels of vinblastine, one solution to increase yields of vincristine is to convert it from vinblastine using laboratory techniques. Other alkaloids in the plant,

including catharanthine and vindoline, can be converted into vinblastine, to increase the supply of this alkaloid. Furthermore, cell cultures of the Madagascar periwinkle have been investigated as another way to increase stocks of both vincristine and vinblastine, although these have not yet been entirely successful.

Humans cannot match the skills of plants to synthesize very complicated chemicals such as vincristine. But we can use our knowledge of chemistry and biotechnology to find other ways to increase yields of much-needed pharmaceutical drugs.

The African bushwillow

Many plants have been explored for their potential to yield useful anticancer drugs, including the African bushwillow (*Combretum caffrum*), a tree in the bushwillow family (Combretaceae) that is native to the Cape Provinces in South Africa. The bark contains chemicals called combretastatins that have shown anticancer effects in laboratory studies. These chemicals may bind to tubulin in cancer cells, so might act in a similar way to vincristine and vinblastine. The combretastatins and their derivatives are undergoing tests in clinical studies to reveal if they may have any place in cancer chemotherapy. One synthetic derivative is fosbretabulin, which reduces the blood flow to tumours. It is under investigation for use in lung, ovarian and thyroid cancers.

RIGHT African bushwillow (*Combretum caffrum*) is a spreading tree that can reach 10 m (33 ft) in height and usually grows along rivers and streams. The fruit is a four-winged nut, and the bark, leaves and roots are traditionally used as a general tonic and for body pain.

THE SPINDLE FAMILY

PLANT:

Gymnosporia rothiana
(Walp.) M.A.Lawson (syn.
Maytenus rothiana (Walp.)
Ramamoorthy)

COMMON NAME(S):

Roth's spike-thorn

FAMILY:

spindle (Celastraceae)

ACTIVE COMPOUND(S):

NATURALLY OCCURRING:

ansamycin macrolide
(maytansine)

SEMI-SYNTHESIZED: DM1,
trastuzumab-DM1
(trastuzumab emtansine)

MEDICINAL USES:

MAIN: cancer chemotherapy

PARTS USED:

stem wood and bark

Maytansine

ABOVE **Maytansine was first reported to occur in species of the spindle family (Celastraceae), although recent studies suggest it may be produced by microbes associated with these plants. A derivative of maytansine, DM1, has been conjugated with a specific antibody to develop a new therapeutic agent for certain cancers.**

In the 1970s, the complex macrolide compound maytansine was discovered in a few plants in the spindle family (Celastraceae), including the seeds of Roth's spike-thorn (*Gymnosporia rothiana*), a small tree native to southwest India. Maytansine was also reported as a constituent of espinheira santa (*Maytenus truncata*), which is in the same family and is native to east Bolivia and Brazil. This shrub was traditionally used in South American medicine systems for its reputed anti-asthmatic, antiseptic and anti-tumour properties, and was also believed to have fertility-regulating effects. In addition, maytansine was discovered in the wood and bark of stems of mock forest spike-thorn (*Putterlickia verrucosa*), which has a native range extending from southern Mozambique to South Africa. From the 1970s, there was much interest in the compound as a potential anticancer drug, as it was found to have potent tumour-inhibiting properties. Maytansine, and its analogue DM1, were discovered to have microtubule-inhibiting effects, which explained their toxicity to cancer cells and provided the basis for drug development.

DYNAMIC DUO

While small molecules from plants were being investigated or developed as potential anticancer drugs, further advances in medical research were providing other new therapeutic strategies to target cancer. This included the use of antibodies that have high affinity for certain types of cancer cells and inhibit their proliferation. For example, the monoclonal antibody trastuzumab is directed to tumour cells that have a protein on their surface produced by the human epidermal growth factor receptor 2 (HER2) gene. The HER2 protein is expressed in about one-third of all breast cancers,

so trastuzumab is used as a therapy in these cancers to inhibit tumour cell proliferation.

With these evolving technologies and advances in cancer therapeutics, the search for new anticancer drugs from plants began to decline. However, a new way for the maytansine derivative DM1 to join this medicines revolution was found. DM1 was conjugated with the antibody trastuzumab to provide a new therapeutic agent called trastuzumab emtansine, which is given to patients with HER2-positive breast cancer.

A little help from friends?

Although the anti-tumour compound maytansine was originally discovered in various members of the spindle family, are these plants really the true sources of maytansine? Recent research suggests that maytansine is instead produced by an endophytic bacterial community associated with the roots of these plants. It appears that, when they synthesize useful compounds, plants might have a little help from other microorganisms.

Indeed, other anticancer drugs have been sourced from bacterial cultures. For example, daunorubicin and doxorubicin are anthracycline chemicals from *Streptomyces* species of bacteria. These drugs interact with DNA and interfere with its replication, so are toxic to tumour cells and are used for a range of different forms of cancer.

TOP AND ABOVE **Roth's spike-thorn (*Gymnosporia rothiana*) is a large shrub from India. Its small flowers (below) are borne in clusters on the stems and are followed by three-lobed fruit (above), which ripen to reddish brown before splitting open to release the small round seeds.**

Signs of the skin

Skin cancer is often caused by ultraviolet radiation from the sun, which damages the DNA in skin cells. Chronic sun damage can cause skin cells to become rough and scaly, in some cases leading to actinic (or solar) keratosis, which can be an early sign of skin cancer. Some drugs have been developed for topical use against this condition, including ingenol mebutate from petty spurge, and masoprocol from the creosote bush.

SAP FOR THE SKIN

PLANT:

Euphorbia peplus L.

COMMON NAME(S):

petty spurge, milkweed

FAMILY:

spurge (Euphorbiaceae)

ACTIVE COMPOUND(S):

NATURALLY OCCURRING:

macrocyclic diterpene ester (ingenol mebutate)

MEDICINAL USES:

MAIN: actinic keratosis

OTHER: warts

PARTS USED:

sap

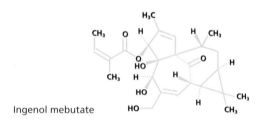

Ingenol mebutate

ABOVE **The sap from the petty spurge (*Euphorbia peplus*) contains ingenol mebutate, which has been developed as a pharmaceutical preparation that is applied to the skin for the condition actinic keratosis.**

Petty spurge is a poisonous plant that has been linked to the deaths of horses, cattle and sheep in New Zealand and Australia. However, the plant's Gaelic name, *lus leighis*, means 'healing herb' and gives a hint of its potential medicinal powers. Indeed, physicians in the Scottish Highlands used milkweed as a remedy for skin cancer for centuries. The plant is also known as wart wort or wart grass, referring to the traditional use of the sap to combat warts. The milky sap was also once used as a purgative and for asthma and catarrh.

Petty spurge has a native range spanning from the Mediterranean to Somalia, and from Europe to the western Himalayas, but it has been introduced to other parts of the world. It is in Australia that the origins of its importance for drug discovery began. Considering the traditional application of the sap for skin cancer, researchers in Australia started to explore the science behind this use in the 1990s. Laboratory studies showed the sap inhibited the proliferation of skin tumour cells more selectively compared to healthy skin cells. The sap was later tested in people with different forms of skin cancer lesions and showed some promising results.

The next challenge was to isolate the main active ingredient. This was revealed to be an extremely complex macrocyclic diterpene ester, named ingenol

mebutate. While its mode of action has not yet been fully elucidated, it can modulate the action of protein kinase C, which is involved in signalling pathways to mediate cancer cell death, and it stimulates the immune response. This historical remedy for skin cancer has provided a new drug, ingenol mebutate, that today is formulated in a gel for the treatment of actinic keratosis.

THE CREOSOTE BUSH

Creosote bush leaves, also known as chaparral, were once used as a remedy for conditions ranging from arthritis, colds and tuberculosis, to cancer. There has been interest in use of the plant against skin cancer, although when taken internally it has been linked with liver toxicity. The bush has a native range stretching from the southwest and south-central United States to Mexico, and has also been studied to determine if any of its chemical constituents might be useful against skin cancer. Masoprocol (also known as mesonordihydroguaiaretic acid) is one such constituent and has been investigated as a topical treatment for actinic keratosis. However, its use has not been widespread as it may cause skin irritation when applied (see also page 97).

BELOW LEFT **Petty spurge (*Euphorbia peplus*) is a small annual of cultivated or waste ground that can grow to 10–30 cm (4–12 in) tall. It produces a milky sap, and its very small flowers are followed by three-lobed fruit.**

RIGHT **Creosote bush (*Larrea tridentata*) is a woody desert shrub whose leaves and young stems produce an aromatic resin. Its five-petalled yellow flowers can measure up to 3 cm (1 in) in diameter.**

Potato family pharmaceuticals

The potato family includes numerous species containing compounds that are glycosides of the steroidal alkaloid solasodine (see page 201). These include solamargine, solasonine and related compounds, which have been shown to cause selective toxicity against skin cancer cells. For this reason, a preparation containing solasodine glycosides – combined with keratolytic salicylic acid (see page 170) and oil from the leaves of tea tree (*Melaleuca alternifolia*; see page 170) in the myrtle family (Myrtaceae), among other ingredients – was developed in Australia as a topical treatment for actinic keratosis.

Food with function

Many plants in our diets have been studied for any protection they may offer against cancer. Those investigated include turmeric, cruciferous vegetables, tomatoes (*Solanum lycopersicum*, syn. *Lycopersicon esculentum*; in the potato family) and some berry fruits. Their chemical constituents have also been explored, to provide an understanding of how they might be useful against cancer.

A GOLDEN SPICE

PLANT:
Curcuma longa L. (syn. *Curcuma domestica* Valeton)

COMMON NAME(S):
turmeric

FAMILY:
ginger (Zingiberaceae)

ACTIVE COMPOUND(S):

NATURALLY OCCURRING:
curcuminoids (curcumin)

MEDICINAL USES:

MAIN: digestive disorders, anticancer

OTHER: anti-inflammatory

PARTS USED:
rhizome

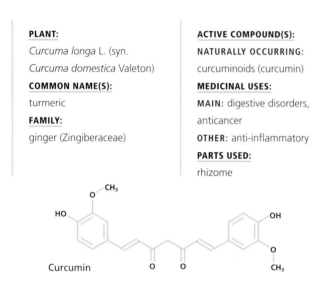

Curcumin

ABOVE **Yellow to orange curcuminoid pigments make up as much as 5 per cent of the spice turmeric (*Curcuma longa*). One of these is curcumin, which has been of interest for its potential anticancer effects.**

Turmeric is a golden-yellow spice that is a popular ingredient of curries and other foods, for its flavour and as a colouring agent. While it is most familiar for its culinary uses, it has also long been renowned in Ayurvedic medicine for its anti-inflammatory and anti-ageing properties. Other uses of turmeric include to aid bile flow, relieve indigestion and counteract liver disorders (see page 106).

Some studies suggest that in countries where people consume more turmeric, the rates of certain cancers appear to be lower. The plant, which is native to India, has therefore been studied extensively for its medicinal effects, including its potential to prevent some forms of cancer. The main biologically active constituents in the rhizome are curcuminoids, including curcumin, demethoxycurcumin and bisdemethoxycurcumin. Most research has focused on curcumin, which in laboratory studies has shown anti-tumour effects through different

mechanisms. For example, curcumin appears to kill cancer cells and to interfere with signalling pathways between cells, preventing cancer cell proliferation.

Free radicals can damage the cells in our body. This leads to what is known as oxidative stress, which has been implicated in diseases such as cancer and heart disease. Curcumin is a powerful antioxidant, so may help reduce oxidative stress and may also enhance our own antioxidant defence systems. Another laboratory study revealed that when curcumin is combined with conventional chemotherapy drugs, the effects might be better than chemotherapy alone. Curcumin has been tested in people with certain forms of cancer, including in combination with chemotherapy drugs. Some studies suggest it may have positive effects, although more clinical studies with humans are needed to determine any usefulness. One of the challenges with curcumin as a component of turmeric in the diet, or as a supplement, is that it is not easily absorbed when taken orally. Some researchers are therefore developing new formulations of curcumin to improve its absorption.

SEEING RED

Carotenoids are plant chemicals that are responsible for the yellow, orange and red colours of some fruits and vegetables, and include *beta*-carotene, which contributes to the bright orange colour of carrots (*Daucus carota* subsp. *sativus*) in the carrot family (Apiaceae). Carotenoids are present in many other vegetables, such as cabbage, but here their colour is masked by the green chlorophyll in the leaves.

There has been much interest in the anticancer effects of the carotenoid lycopene, which occurs in tomatoes. Lycopene can scavenge free radicals, so it may help to protect our cells – and cell components, such as DNA – from damage caused by oxidative stress. Studies that associate diet and disease risk indicate that diets rich in tomatoes may reduce lung cancer rates, while other studies reveal that a high intake of tomatoes may reduce the risk of prostate cancer.

A plate of prevention?

The World Health Organization has concluded that diets high in fruits and vegetables may have a protective effect against many cancers. Scientific studies have revealed that anthocyanins, responsible for the blue colour of berry fruit such as blueberries (*Vaccinium* spp.) and black raspberries (*Rubus occidentalis*) in the rose family (Rosaceae), can act by different mechanisms that might help prevent some forms of cancer.

Other studies indicate that a high intake of cruciferous vegetables such as cabbage and broccoli, which are rich in glucosinolate chemicals, may reduce the risk of certain cancers. Garlic (*Allium sativum*) in the amaryllis family (Amaryllidaceae) (see also page 49) has also been of interest for its potential anticancer properties. What we put on our plates can be as important for our health as the medicines we take.

LEFT Turmeric (*Curcuma longa*) flowers are sterile (they do not produce viable seeds) and form in the lower bracts of congested heads, which are borne on short, 10–15 cm-long (4–6 in) stalks. Plants are propagated by division of the rhizome.

Glossary

acetylcholine neurotransmitter that mediates the transmission of nerve impulses by binding to muscarinic and/or nicotinic receptors in the nervous system. Degraded by the enzyme acetylcholinesterase.

adaptogen substance that aids non-specific resistance to adverse influences, such as stress and disease.

adrenaline chemical messenger secreted by the adrenal glands and certain nerve cells; also known as epinephrine.

alkaloid nitrogen-containing compound; alkaloids often have pronounced physiological effects in humans.

analgesic reduces or relieves the sensation of pain.

annual plant that completes its life cycle in one growing year.

antagonize have an opposite effect to, or inhibit the activity of, another substance.

anti-arrhythmic regulates or manages arrhythmias (variation in the regular rhythm of heartbeats).

anticholinergic compound that inhibits the action of acetylcholine on cholinergic (muscarinic and/or nicotinic) receptors in the nervous system.

antidote substance that counteracts the effects of a poison.

antipyretic reduces fever.

aril outer seed covering, often fleshy.

arrhythmia abnormal heart rhythm.

benign prostatic hyperplasia enlargement of the prostate.

biennial plant that usually completes its life cycle in two growing years, producing foliage only in the first year, then flowers and fruits in the second.

blood–brain barrier physiological barrier (semi-permeable membrane) that separates the blood from the brain and cerebrospinal fluid within the brain and the spinal cord.

carcinogenic having the potential to cause cancer.

cardiac pertaining to the heart.

cardiac glycoside plant compound that has a steroidal-type structure; cardiac glycosides may increase the force of the contraction of the heart.

central nervous system the brain and spinal cord in higher animals.

cholera infectious disease caused by the bacterium *Vibrio cholerae*; involves symptoms such as diarrhoea.

cladode flattened and expanded stem or branch that functions as a leaf.

clinical study (clinical trial) controlled study or experiment that involves a defined set of human subjects that receive a form of medical intervention (such as a medicine) to obtain scientifically valid information about the efficacy and safety of that medical intervention.

cultivar plant variety usually produced by selective breeding in cultivation.

decoction preparation that involves boiling a herbal drug in water prior to straining.

dermatitis inflammation of the skin.

diuretic promotes production and excretion of urine.

dopamine neurotransmitter that occurs in the nervous system.

dysentery disease caused by an infection; which produces symptoms of watery diarrhoea and fever.

dyspepsia disorder of the stomach and gut that often involves pain, nausea and indigestion.

emetic compound that causes vomiting.

enzyme protein that acts as a catalyst for chemical reactions.

evergreen plant that retains leaves for more than one growing season.

expectorant aids the removal of secretions (sputum/mucus) from the air passages, especially to relieve coughs.

glycoside non-sugar compound with one or more sugar molecules attached.

haemorrhoids enlarged blood vessels in or around the anus and rectum; also known as piles.

hallucinogenic causing an altered mental state, distorting perceptions, with visual or auditory sensations.

hepatitis inflammation of the liver.

histamine chemical mediator in the body with different functions, especially in allergic reactions.

hypertension high blood pressure.

infusion preparation obtained by steeping a herbal drug in water.

jaundice increased levels of bile pigments in blood, causing a yellowish discoloration in some body tissues, including skin, and in excretions.

latex viscous sap or fluid produced by some plants.

muscarinic receptor type of cholinergic receptor on nerve cells that transmits signals after activation by acetylcholine, but also by muscarine and similar compounds (cf. nicotinic receptor). Inhibited by antimuscarinic compounds.

neuralgia nerve pain.

neuron nerve cell.

neurotransmitter chemical substance released by nerve cells to transmit signals to other cells.

nicotinic receptor type of cholinergic receptor on nerve cells that transmits signals after activation by acetylcholine, but also by nicotine and similar compounds (cf. muscarinic receptor).

noradrenaline chemical messenger that is a precursor of adrenaline in the adrenal glands and is present in the brain; also known as norepinephrine.

oedema swelling caused by the collection of fluid in body tissues or cavities.

opiate narcotic substance that can be prepared from opium.

opioid compound with the same effects as opium or its derivatives, e.g. morphine.

pathogen any microorganism, virus or other biological substance that causes disease.

perennial plant that lives for two years or longer and, once mature, flowers annually.

peripheral nervous system the nervous system outside the brain and spinal cord in higher animals, including nerves responsible for the five senses, movement, and sending other signals between the body and brain.

placebo inactive substance often included in clinical trials to compare with a drug substance under study.

post-herpetic neuralgia nerve pain that occurs following infection with the virus that causes shingles.

poultice preparation applied to the skin formed by wetting an absorbent substance with fluids that may be medicated.

psoriasis condition that involves inflamed and scaling lesions of the skin.

purgative laxative, usually with pronounced activity.

rheumatism various conditions associated with joint or muscle pain.

saponin plant-derived compound with a water-soluble part (of sugars) attached to a lipid-soluble core structure, causing it to foam in water.

spasmolytic prevents or alleviates muscle spasms (cramps).

steroid compound with a structure that chemically resembles cholesterol and that has certain physiological activities, e.g. cardiac glycosides and sex hormones.

stigma and style female part of a flower.

stomachic substance that improves appetite or digestion.

terpene compound derived from two or more isoprene units.

topical the delivery of a medicinal substance by application to the exterior of the body

toxin poisonous substance produced by a biological organism, e.g. a microbe, animal or plant.

tuberculosis disease caused by infection with the bacterium *Mycobacterium tuberculosis*, which can affect almost any tissue or organ of the body, e.g. the lungs.

vitiligo condition that involves the appearance of non-pigmented areas on skin.

Further reading

Deni Bown, *The Royal Horticultural Society Encyclopedia of Herbs*. Dorling Kindersley, London, 2014.

Maarten Christenhusz, Michael Fay and Mark Chase, *Plants of the World: an illustrated encyclopedia of vascular plant families*. Royal Botanic Gardens, Kew, and University of Chicago Press, Chicago, IL, 2017.

Elizabeth A. Dauncey and Sonny Larsson, *Plants That Kill*. Royal Botanic Gardens, Kew, 2018.

Sarah E. Edwards, Inês da Costa Rocha, Elizabeth M. Williamson and Michael Heinrich, *Phytopharmacy: an Evidence-Based Guide to Herbal Medicinal Products*. Wiley, Chichester, 2015.

Steven Foster and James A. Duke, *Peterson Field Guide to Medicinal Plants and Herbs of Eastern and Central North America*, 3rd edn. Houghton Mifflin, Boston, 2014.

Steven Foster and Rebecca L. Johnson, *Desk Reference to Nature's Medicine*. National Geographic Society, Washington, DC, 2006.

Michael Heinrich, Joanne Barnes, Simon Gibbons and Elizabeth M. Williamson, *Fundamentals of Pharmacognosy and Phytotherapy*, 2nd edn. Elsevier, London, 2012.

Christine Leon and Lin Yu-Lin, *Chinese Medicinal Plants, Herbal Drugs and Substitutes: An Identification Guide*. Royal Botanic Gardens, Kew, 2017.

Pharmaceutical Press Editorial, *Herbal Medicines*, 4th edn. Pharmaceutical Press, London, 2013.

Gunnar Samuelsson and Lars Bohlin, *Drugs of Natural Origin*, 6th edn. Swedish Academy of Pharmaceutical Sciences, Stockholm, Sweden, 2009.

Monique Simmonds, Melanie-Jayne Howes and Jason Irving, *The Gardener's Companion to Medicinal Plants: an A–Z of healing plants and home remedies*. Frances Lincoln, London, and Royal Botanic Gardens, Kew, 2016.

Ben-Erik van Wyk and Michael Wink, *Phytomedicines, Herbal Drugs, and Poisons*. University of Chicago Press, Chicago, IL, and London, and Royal Botanic Gardens, Kew, 2014.

Online resources

Bob Allkin et al., 'Useful plants – medicines'. In: Kathy Willis (ed.), *State of the World's Plants 2017*, Royal Botanic Gardens, Kew. Available online at https://stateoftheworldsplants.org

Medicinal Plant Names Services Portal, Royal Botanic Gardens, Kew, Version 8. http://kew.org/mpns

Plants of the World Online, Royal Botanic Gardens, Kew. http://www.plantsoftheworldonline.org

Index

acanthus family (Acanthaceae) 119, 127
acetylsalicylic acid see aspirin
Acmella oleracea 77
Acokanthera schimperi 37, 39, 108–9
Actaea spp. 62–3, 194–5; *A. dahurica* 63; *A. racemosa* 63, 185, 194–5
actein, 27-deoxyacetin 194
Adansonia digitata 31
Adhatoda vasica see *Justicia adhatoda*
Adoxaceae see moschatel family
aescin 50–1
Aesculus hippocastanum 37, 50
African medicine, introduction 12, 108–9
African plum/prune 110–11
Afzelia spp. 97
Agave spp. 22, 163; *A. sisalana* 28, 125, 160, 162–3; *A. tequilana* 163
agave, blue/tequila 163
ageing 65, 102, 167, 216
agnus-castus 192
ajmaline 40, 47
alkaloids 19, 21, 22, 29, 31, 33, 40, 47, 55, 68–9, 73, 77, 108–9, 112, 118, 120–1, 125, 135, 138, 151, 161, 200–1, 205, 207; bisbenzylisoquinoline 152–3; bisindole 210–11; imidazole 178–9; indole 30–1, 36, 40, 47, 61, 69, 72, 75, 153, 155, 179, 196; isoquinoline 47; isoquinoline-derived 70, 156; morphinan 140–1, 155; piperidine 29; phenethylisoquinoline 148; purine 66–7; pyridine 68, 98; pyrrolidine 68; pyrrolizidine 185; quinazoline 119; quinoline 41, 208–9; quinolizidine 31, 40, 69; steroidal 30, 33, 46, 215; tropane 33, 72, 76–7, 86–7, 91, 156, 180–1
alkylamides 126–7
allergies 91, 117–18, 124–5
allicin, alliin 49
Allium spp. 147; *A. sativum* 37, 49, 147, 217
aloe, Réunion 165
aloe family see asphodel family
Aloe spp., *A. macra*, *A. purpurea* 165
Aloe vera 19, 160, 162, 164–5
Amanita spp., *A. phalloides*, *A. virosa* 105
amaryllis family (Amaryllidaceae) 37, 55, 70, 109, 147, 217
amatoxins, amanitin 102, 104–5
ambroxol 119
amino acids 31, 154
amiodarone 36, 41, 124
Ammi visnaga see *Visnaga daucoides*
ammi, greater (*Ammi majus*) 159–60, 168–9
anabasine 69
anaesthesia 21, 55, 76–7, 87
Ananas comosus 167
Andira araroba see *Vataireopsis araroba*
andrographis (*Andrographis paniculata*) 126–7

andrographolides 127
anethole, *trans-* 131
Anethum graveolens 29
angel's trumpets 87
angina/angina pectoris 41, 43, 46–7, 124
anisatin 131
anise 29
Annonaceae see soursop family
ansamycin macrolide 212
anthocyanins 18–19, 217
anthracyclines 212
anthralin see dithranol
anthraquinones 18; glycosides 93
anthrone derivatives 31, 169
anti-inflammatories 21, 49, 51, 65, 75, 82, 88, 123, 127–8, 142–3, 146, 160–5, 167, 169, 171, 173, 187, 191, 205, 216
Apiaceae see carrot family
Apium graveolens 29
Apocynaceae see dogbane family
apomorphine 139, 155
Aquifoliaceae see holly family
Araliaceae see ivy family
araroba 31, 160, 169
Areca catechu 80, 98–9, 129
Arecaceae see palm family
arecoline 80, 98; hydrobromide 98
Aristolochia fangchi 17
Aristolochiaceae see birthwort family
aristolochic acid 17
arnica (*Arnica montana*) 160–1, 167
artemether 134–5
Artemisia annua 23, 63, 119, 134; *A. cina* see *Seriphidium cinum*; *A. maritima* 99
artemisinin 23, 119, 134–5
artesunate 134
arthritis (rheumatoid/osteoarthritis) 22, 50–1, 57–9, 62, 82, 93, 114, 122–3, 129, 138, 143–7, 165, 167, 177, 180–1, 187, 191, 195, 215
artichoke, globe 105
ashwagandha 45, 65
asiatic acid, asiaticoside 166
Aspalanthus linearis 108
asparagus 28; fern, 45
asparagus family (Asparagaceae) 13, 22, 28, 37, 39, 45, 81, 125, 160, 162
Asparagus spp., *A. officinalis* 28; *A. racemosus* 45
asphodel family (Asphodelaceae) 19, 81, 128, 160, 164
Aspidium filix-mas see *Dryopteris filix-mas*
aspirin (acetylsalicylic acid) 21, 23–4, 37, 48–9, 138, 140, 142–3, 146
Asteraceae see daisy family
asthma 41, 57, 66–7, 69, 74, 108, 117–18, 120–1, 124–5, 144, 156, 212, 214
atracurium 152–3
Atropa bella-donna 20, 72, 86, 156, 161, 180
atropine 20, 72, 86, 156, 178, 180–1
Atuna racemosa ssp. *racemosa* 128

aubergine 33
autumn crocus family 138, 148
Avena sativa 164–5; *A. sterilis* 164
avenanthramides 165
axi 151
Ayurvedic medicine, introduction 14, 44–5

Bacopa monnieri 167
bakuchi 168
balsam: Peru 160–1, 176–7; tolu 160, 176
balsamic esters 31, 176–7
banana, Ethiopian 109
banana family 109
Banksia spp. 129
baobab tree 31
barbasco 184, 186–7
barberry family 161, 172, 185, 195, 200, 202, 204
barley 55, 77
barrenwort 185, 197
bean: broad 154; Calabar 21, 55, 70, 72–3, 108, 161, 178–9; faba 154; soya (soybean) 195; velvet 154–5
bean family see legume family
beech 96–7; European 96–7; Japanese 96
beech family 96
beleric/belleric myrobalan 45, 85
belladonna 20, 72, 86, 156, 161, 178, 180–1
bellflower family 69
belotecan 209
benzatropine (benztropine) 156
benzoin 175
benzyl benzoate 161, 176–7
Berberidaceae see barberry family
bergamot, wild 205
bethroot 195
Betula spp. 22; *B. lenta* 143; *B. pendula* 23
Betulaceae see birch family
betulinic acid 22
biguanides 112–13
bibhitaki 45
bilobalide 74–5
birch 22; silver 23; sweet 143
birch family 22, 143
birthroot 195
birthwort, Chinese 17
birthwort family 17
bis-sesquiterpene 188
bisdemethoxycurcumin 216
black-eyed Susans 127
blood clots 37, 48–9, 74–5
blood pressure, high/hypertension 36, 40, 42–3, 46–7, 61
blueberries 217
Boophone disticha 109
borage (*Borago officinalis*) 185
borage family (Boraginaceae) 185
Boswellia sacra 84–5
brahmi 167
Brassica oleracea 147, 201
Brassicaceae see cabbage family
broccoli 147, 201, 217
bromelain 167
bromhexine 119
bromocriptine 155
broom 40–1; butcher's 28, 37, 50–1
buckthorn 93; alder 18
buckthorn family 18, 81, 93
bufadienolides 28
buformin 113

bugbane 62, 194; black 194; dahurian 63
buprenorphine 141
Burseraceae see frankincense and myrrh family
bushwillow, African 211
bushwillow family 45, 85, 211
buttercup family 62, 185, 194, 204

cabazitaxel 206–7
cabbage 147, 201, 216–17
cabbage family 147, 185, 201
cabeza de negro 184–7
caffeic acid 126
caffeine 31, 55, 64, 66–7, 77, 108, 125
Calamus draco 81
Calendula officinalis 191
caltrop family 97, 118, 122, 201, 205
Camellia sinensis 25, 55, 62, 95, 147, 172
Campanulaceae see bellflower family
camphor 121, 138, 145
camphor tree 121, 138, 145
Camptotheca acuminata 21, 201, 208–9
camptothecin 21, 201–2, 208–9
Cananga odorata 128, 175
cancer 19, 21–3, 30, 64, 68, 80, 107, 113, 132–3, 146, 148, 172–3, 199–203, 206–17
cannabidiol 153
cannabinoids 22, 153
cannabis (*Cannabis sativa*) 80, 82–3, 152–3
cannabis family (Cannabaceae) 58, 80, 152
Caprifoliaceae see honeysuckle family
capsaicin 138, 140, 144
capsaicinoids 144–5
Capsicum spp. 151; *C. annuum* 33, 138–9, 144, 151
Carapichea ipecacuanha 154
caraway 29
carbenoxolone sodium 80, 88–9
cardamom 45
cardiac glycosides 21, 28, 30, 38–9
Carduus marianus see *Silybum marianum*
Carica papaya 167
Caricaceae see papaya family
carotene, *beta-* 18, 216
carotenoids 18, 74, 216
carrot family 18, 29, 36, 45, 118, 124, 143, 160, 166, 168, 191, 216
carrots 18, 29, 216
Carthamus tinctorius 115
Carum carvi 29
cascara sagrada 93
cascarosides 93
cassava 33
Cassia acutifolia, *C. angustifolia*, *C. senna* see *Senna alexandrina*
casticin 192
castor oil plant 12–13, 33, 177
catbrier family 163, 186
catechins 25, 42, 94, 173
catharanthine 211
Catharanthus roseus 22, 199, 201, 210
Caulophyllum thalictroides 195
Celastraceae see spindle family
celery 29
Centella asiatica 29, 160, 166–7
Central American medicine, introduction 150–1

Cephalotaxus harringtonia 22, 207
ceremonies and rituals 54, 56–7, 73, 108, 128–9, 140, 150–1, 180, 186, 194
chaparral 97, 205, 215
charantin 114
chaste tree 192
chasteberry 183, 185, 192–3
chhataya 168
Chinese medicine, introduction 16–17, 62–3
chloroquine 135
chocolate 31, 151
Chondrodendron tomentosum 139, 152
chrysarobin 169
Chrysobalanaceae see cocoplum family
Cicuta virosa 29
cicutoxin 29
Cimicifuga dahurica see *Actaea dahurica*; *C. racemosa* see *A. racemosa*
cimicifugosides 194
cinchona 20, 94, 119, 134–5; yellow 135
Cinchona spp. 20, 36, 41, 94, 119, 135
Cinnamomum camphora 121, 138, 145; *C. verum* 45
cinnamon, Ceylon 45
circulatory system 28, 35–43, 46–51, 53, 74, 95, 123, 144, 167, 176–7, 179, 181
Cissampelos pareira 153
citrus family 98, 161, 178
Claviceps purpurea 139, 155
Clavicipitaceae 139, 155
clover, king's 48; red 195; sweet 31, 37, 48; yellow sweet 37, 48
cloves 76–7, 174
clubmoss, toothed 73
clubmoss family 73
coca 21, 33, 55, 68, 76–7
coca family 21, 33, 55
cocaine 21, 33, 55, 68–9, 76–7
cocoa 31
cocoplum family 128
codeine 20, 47, 96, 140–1
Coffea spp. 55, 108; *C. arabica* 66–7, 125; *C. canephora*, *C. liberica* 67
coffee 55, 66–7, 108; arabica 66–7, 125; liberica 67; robusta 67
coffee family 20, 36, 55, 66–7, 108, 119, 125, 154, 185, 196, 209
cohosh, black 63, 185, 194–5; blue 195
cohuanenepilli 151
cola (*Cola acuminata*; *C. nitida*) 31, 55, 66, 77
Colchicaceae see autumn crocus family
colchicine 138, 148
Colchicum autumnale 138, 148
Coleus barbatus, *C. forskohlii* 32
Combretaceae see bushwillow family
combretastatins 211
Combretum caffrum 211
Commelina communis 103, 114
Commelinaceae see spiderwort family
Commiphora myrrha 84
coneflower 117, 126–7; narrow-leaf 126; pale 126; purple 118, 126–7, 205; yellow 127
conessi, conessine 30
coniine 29
Conium maculatum 29

conservation, wild collection and sustainability 22, 26–7, 65–7, 71, 74, 87, 91, 93, 95, 111, 123, 125, 128, 147, 163, 165, 187, 191, 197, 203, 206–8

constipation (laxatives) 13, 18, 31, 45, 79, 81, 84, 90, 92–3, 164, 204

Convallaria spp. 28; *C. majalis* 28, 39

coriander (*Coriandrum sativum*) 29

corkwood 90–1

Cornaceae see dogwood family

Cornus spp. 22; *C. florida* 23

corticosteroids 28, 125, 160, 163, 187

cortisone 186

Corynanthe johimbe 108, 185, 196

cotton 31, 184, 188–9; Levant 188–9; Mexican/upland 188

cough 47, 51, 62, 89, 96, 105, 117–18, 121–2, 127, 131, 140–1, 143–4, 148–9, 156, 205

coumarins 31, 48

coumestans 195

Crataegus laevigata 42; *C. monogyna* 37, 42

creosol, cresol 96

creosote: bush 97, 201, 205, 214–15; beechwood 81, 94, 97; coal tar 97; wood 96–7

creosotum ligni 96

Crocus sativus 60–1

crocus: autumn 138, 148–9; saffron 60–1

crofelemer 25, 81, 94–6

cromolyn sodium see sodium cromoglicate

croton, purging 133

Croton spp. 33, 94; *C. lechleri* 81, 94–5; *C. tiglium* 133

cruciferous vegetables 201, 216–17

Cryptocarya woodii 12

Cryptomeria japonica 97

cucumber family (Cucurbitaceae) 114

Cullen corylifolium 31, 168

Culver's root 205

cumin (*Cuminum cyminum*) 45

Culpeper, Nicholas 42–3, 59, 75, 143, 149, 191

curare: vine 139, 152–3; calabash 153

Curcuma longa (*C. domestica*) 45, 102, 107, 147, 201, 216–17

curcumin 106–7, 216

curcuminoids 216

cyanogenic glycosides 33

cyclopeptides 104

Cynara cardunculus var. *scolymus* 105

Cyperaceae see sedge family

Cyperus papyrus 13

cytisine 69

Cytisus scoparius 40–1

Daemonorops draco see *Calamus draco*

daffodils 70–1

dahlia (*Dahlia coccinea*) 151

daisy: electric 77; pyrethrum 55

daisy family 23, 63, 77, 80, 99, 102, 104–5, 115, 118–19, 126, 134, 151, 160, 191, 205

dantron (danthron; 1,8-dihydroxyanthraquinone) 92–3

Datura ceratocaula, D. innoxia 151; *D. stramonium* (*D. tatula*) 139, 156–7

Daucus carota ssp. *sativus* 18, 216

daunorubicin 212

dayflower, Asiatic/common 103, 114

10-deacetylbaccatin III 207

dead man's bells 39

death cap 105

death's flower 71

dementia, Alzheimer's disease 21–3, 30, 55, 64, 67, 70–5, 167, 179

demethoxycurcumin 216

1-deoxynojirimycin (1-DNJ) 103, 114

depression 40, 47, 58, 60–1, 108

dermatitis 162–3

desaspidin 99

deserpidine 47

destroying angel 104–5

devil's: breath 87; claw 109, 138, 146–7, 153, 203

devildoer 153

dextromethorphan 141

diabetes 45, 64, 69, 101, 103, 112–15, 167, 210

dianthrone glycosides 31, 92

diarrhoea 30, 33, 51, 66, 79, 81, 90, 94–7, 128, 141, 181, 205

dicoumarol 48

diflomotecan 209

digestive system 13–14, 29, 33, 45, 47, 58–9, 64, 69, 77, 79–83, 86–99, 103, 105–6, 109–10, 115, 125, 128–9, 132, 144, 146–7, 156, 165, 173, 175, 191, 204–5, 214, 216

Digitalis spp. 21, 38–9; *D. lanata* 36, 38; *D. purpurea* 8, 35–6, 39

digoxin 22, 36, 38–9

dihydrocodeine 141

dill 29

dimethylbiguanide 112

dioscin 186

Dioscorea spp. 125, 163, 184; *D. composita* 184, 186–7; *D. deltoidea* 125; *D. mexicana* 184–6; *D. tokoro* 186; *D. villosa* 186–7

Dioscoreaceae see yam family

Dioscorides, Pedanius 14–15, 149, 187, 190–1, 193

diosgenin 125, 163, 184–7, 195

dipterocarp family (Dipterocarpaceae) 97

diterpenes 32–3, 74, 127, 132–3, 192, 206, 214

dithranol (anthralin) 168–9

DM1 22, 212

docetaxel 206–7

dock, curly/yellow 168–9

dock family see knotweed family

doctrine of signatures 15, 40, 51, 65, 105, 142

dogbane family 22, 30, 36, 39–40, 61, 69, 75, 108–9, 201, 209–10

dogwood family 22

dogwoods 22; flowering 23

doxorubicin 212

Dracaena draco 81

dragon's blood 81, 94–5

Drimia maritima 13

dronabinol 153

Dryopteris campyloptera (*D. austriaca*), *D. filix-mas* 99

Duboisia spp. 90–1

dysentery 30, 81, 96, 99, 156

East Asian medicine, introduction 62–3

echinacea 117–18, 126–7, 205

Echinacea sp. 117; *E. angustifolia, E. pallida* 126; *E. paradoxa* 127; *E. purpurea* 118, 126–7, 205

eclectic medicine, introduction 204–5

eczema 39, 129, 163, 169

elder 126–7

Elettaria cardamomum 45

Eleutherococcus senticosus 65

eleutherosides 65

emetine 156

emodin anthrone 93

Ensete ventricosum 109

ephedra 118, 120–1, 125; Chinese 63, 118, 120–1

ephedra family (Ephedraceae) 63, 118, 120, 205

Ephedra spp. 120, 205; *E. nevadensis* 118, 120–1; *E. sinica* 63, 118, 120–1

ephedrine, norephedrine, pseudoephedrine 118, 120–1, 125, 205

epicatechin 42

epigallocatechin gallate 173

Epimedium spp. 185, 197

ergometrine, ergotamine 155

ergot 139, 155

Ericaceae see heather family

Erythroxylaceae see coca family

Erythroxylum coca 21, 33, 55, 76

eserine see physostigmine

espinheira santa 212

essential (volatile) oils 19, 29, 32, 58, 77, 90, 118, 121–2, 161, 170, 175, 177

estradiol, estrone 187

etoposide 201–3

eucalyptus 118, 121, 128–9

Eucalyptus spp. 118, 128; *E. globulus* 121

eugenol 77

Euphorbia peplus 22, 201, 214–15

Euphorbiaceae see spurge family

European medicine, introduction 12, 14, 190–1

evening primrose 185, 193

evening primrose family 185

exatecan 209

eyeball plant 77

eyes 20, 77, 143, 153, 159–61, 178–80

Fabaceae see legume family

Fagaceae see beech family

Fagus spp. 96–7

familial Mediterranean fever 148

fang ji 17; *guang* 17

fatigue 64, 66, 98, 129–30, 167, 195

fatty acids 110–11, 188–9, 193

fennel 29

fenugreek 187

fern: male 99; mountain wood 99

feverfew 191

fig (*Ficus carica*) 84

Filipendula ulmaria 49, 138, 142–3, 171

flavolignans 104

flavonoids 18–19, 31, 42, 61, 74–5, 146, 192, 194–5, 197

flumamine 113

Foeniculum vulgare 29

formononetin 194

forskohlii 32

forskolin 32

forsythia (*Forsythia* spp.) 62

fosbretabulin 211

fountain bush 31, 168–9

fourstamen stephania 17

foxgloves 21–2, 38–9; Grecian 38; purple 8, 35–6, 39; woolly 36, 38–9

Frangula alnus 18; *F. purshiana* 93

frankincense 84–5

frankincense and myrrh family 84

Fuchsia spp. 191

fungi (fungus, mushroom) 8, 19, 139, 155, 160, 170–1

furanochromones 29, 41, 124

furanocoumarins 29, 31

galantamine 55, 70, 71

Galanthus spp. 55; *G. nivalis* 3, 53, 70; *G. woronowii* 70

Galega officinalis 31, 103, 112–13

galegine 112–13

Galen of Pergamon 14–15, 85

Galerina spp. 105

garlic 37, 49, 147, 217

Gaultheria procumbens 143

geraniol 205

Gerard, John 51, 59, 75, 191

ghanera 209

gimatecan 209

ginger 45, 80, 82–3, 106, 216

ginger family 45, 80, 82, 102, 147, 201

gingerols 82

ginkgo (*Ginkgo biloba*) 24, 55, 70, 74–5

ginkgo family (Ginkgoaceae) 55, 74

ginkgolides 74–5

ginseng 15, 54, 62, 64–5, 106; American 65; Indian 33, 45, 65; Peruvian 185; Siberian 65

ginsenosides 64–5

glaucoma 161, 178–9

Gloriosa superba 148–9

glucosinolates 217

Glycine max 195

glycyrrhetic acid 88

Glycyrrhiza glabra 13, 80, 88–9; *G. inflata* 88; *G. uralensis* 88–9

glycyrrhizic acid (glycyrrhizin) 80, 88–9

Goa powder 160, 168–9

goat's rue 31, 103, 112–13, 197

golden rain 69

goldenseal 204

gomisins 106

Gossypium spp. 31, 184, 188–9

gossypol 184, 188–9

gotu kola 29, 160, 162, 166–7

gourd, bitter 103, 112, 114–15

gout 114, 148–9, 165, 191

gramine 55, 77

grapple plant 146

grass family 55, 129, 155, 164

greenbrier 163

guaiac 118, 122

guaiazulene 122

guaifenesin (guaiacol glyceryl ether) 118, 122

guanidine 31, 112

guaraná 66–7

gum Benjamin tree 175

gum trees 121, 128–9; blue 121

gymnosperm 121

Gymnosporia rothiana 212–13

Hamamelidaceae see witch hazel family

Hamamelis spp. 62; *H. virginiana* 204–5

happy tree 21, 201–2, 208–9

harakeke 128

harpagogenin, harpagoside 146

Harpagophytum procumbens 109, 138, 146–7

hawthorn 37–8, 42–3

hayatin 153

headache, migraine 15, 51, 57–9, 66–7, 90–1, 108, 143, 145, 155, 175, 191, 204

heart conditions 19, 21, 28, 30, 36–43, 46–9, 66, 121, 124, 142, 181, 216

heather family 143

hecogenin 125, 162–3

helenalin 167

hellebore: American/green 46–7; European/white/white false 36, 46–7

Helminthotheca echioides (*Helminthia echioides*) 80

hemlocks 29

henbane 79, 86, 91; black 80, 86–7, 156; Egyptian 13, 86–7; Japanese 86; Russian 80, 86

Heracleum candicans 168; *H. mantegazzianum* 29

herb bennett 59

heroin 141

Hevea brasiliensis 95

hibiscus (*Hibiscus* spp.) 31

Hibiscus sabdariffa 18–19

Hippocrates of Kos 14, 140, 143

hogweed, giant 29

Holarrhena pubescens (*H. antidysenterica*) 30

holly, box 51

holly family 66

holywood 118, 122

Homalanthus nutans 119, 132–3

homoharringtonine 207

honeysuckle family 54, 58

hops 56, 58–9

Hordeum vulgare 55, 77

horny goat weed 185, 196–7

horse chestnut 37, 50–1

Humulus lupulus 56, 59

Huperzia serrata 73

huperzine A 73

hyacinth family see asparagus family

Hydrastis canadensis 204

hydroxyanthracenes 81

hyoscine 80, 86–7, 91, 156; hydrobromide 86–7

hyoscyamine 86, 139, 156, 180

Hyoscyamus spp. 86, 91; *H. muticus* 13, 86–7; *H. niger* 79–80, 86–7, 156

hyperforin 60–1

Hypericaceae see St John's wort family

hypericin 60–1

Hypericum perforatum 11, 18, 54, 60

Ibn Sina (Avicenna) 84–5, 190

iboga 69, 108

ibogaine 69

Icacinaceae see white pear family

icariin 197

ifiifi 128

Ilex paraguariensis 66

Illicium spp. 119, 131; *I. anisatum* 131; *I. verum* 119, 130

iminosugars 103, 114

Indian coleus 32

Indian medicine, introduction 14, 44–5

Indian pink family 153

indigestion 48, 82, 91, 144, 216

infections 12, 19, 31–2, 56–7, 59, 62, 64, 81–2, 88, 95–7, 103, 110, 114–15, 117–19, 121–3, 126–7, 131, 147, 152, 159–61, 165, 170–3, 175–6, 179, 195, 205, 212, 215–16
influenza/flu 66, 113, 117–19, 126–7, 130–1, 170
ingenol mebutate 22, 33, 214–15
insect bites and stings 60, 95, 160, 163, 171, 191
insomnia/sleep 21, 32, 54, 56, 58–60, 165, 175, 194
insulin 103, 112–15
Intsia spp. see *Afzelia* spp.
ipecacuanha 154, 156
Iridaceae see iris family
iridoids 58; glycosides 146
irinotecan 208–9
iris family 60
irritable bowel syndrome 32, 45, 81, 90
isothiocyanates 147
ispaghula 81
ivy family 62, 64–5

jaborandi 98, 179; Maranham 161, 178–9; Paraguay/Pernambuco 179
Jamestown weed 156
Jamu medicine, introduction 175
Japanese sugi pine 97
jiang huang 106
jimsonweed 156–7
jus de bissap 19
Justicia adhatoda 119

kalmegh 126–7
Kampo medicine, introduction 16–17, 62–3
karenitecin 209
kava 54, 56–7, 128
kava lactones (kava pyrones), kawain (kavain), 56
kava-kava 56
kawakawa 129
kenguel 105
khella 36, 41, 118, 124–5
khellin 36, 41, 118, 124–5
Kindia gangan 67
knotweed family 18, 81, 168
kola nuts 55, 66–7
kombe 39, 109
Kupeantha spp. 67

Labiatae see mint family
laburnum (*Laburnum anagyroides*) 69
Lamiaceae see mint family
lanatosides 38–9
Larrea tridentata 97, 201, 205, 215
laudanum 140
laurel, Cape 12
laurel family (Lauraceae) 12, 45, 121, 145
lauric acid 110
lavender (*Lavandula angustifolia*) 56–9, 177
lectins 31
legume family 13, 21, 31, 37, 40, 48, 55, 69, 72, 80–1, 88, 92, 97, 103, 108, 112, 139, 154, 160–1, 168, 176, 185, 187, 195
Leguminosae see legume family
lemon balm 56, 58–9, 75
Leonurus cardiaca 43
Lepidium meyenii 184–5
Lepiota spp. 105
Leptandra virginica 205
leptandrin 205
Leucojum spp. 70–1
levodopa (L-DOPA) 31, 139, 154

lidocaine 77
lignans 104, 106, 111, 172–3, 195, 202–3
lignum vitae 122–3; holywood 118, 122; roughbark 118, 122
lilac, French 112
lily-of-the-valley 28, 39
lily: flame/glory 148–9; Loddon 71
lime tree/linden tree 31
linamarin 33
linolenic acid, *gamma*- (GLA) 185, 192–3
lion's tail 43
liquorice 13, 80, 88–9; Chinese 88; common 13, 80, 88–9; Manchurian 88–9
liver disorders 40, 57, 89, 101–2, 104–7, 119, 132–3, 204, 215–16
Lobelia inflata 69
lobeline 69
Loganiaceae see Indian pink family
lorajmine 40
lotus, blue/Egyptian 85
lus leighis 214
lycopene 33, 216
Lycopersicon esculentum see *Solanum lycopersicum*
Lycopodiaceae see clubmoss family
Lycopodium serratum see *Huperzia serrata*
Lythraceae see pomegranate family

ma huang 120
maca 184–5
mace 174–5
Macropiper excelsum 129; *M. methysticum* 54, 56, 128
madecassic acid, madecassoside 166
magnolia (*Magnolia officinalis*) 62
magnolia family (Magnoliaceae) 62
magnolia vine 62; Chinese 102, 104, 106–7
magnolia vine family 62, 102, 119, 130
maidenhair tree 74
makita 128
Malabar nut 119
malaria 20, 23, 30, 41, 94, 113, 117, 119, 134–5, 142, 149, 188
Malesian medicine, introduction 174–5
mallow (*Malva sylvestris*) 31
mallow family (Malvaceae) 18, 31, 55, 151, 184, 188
mamala 119, 132–3
mandrake (*Mandragora officinarum*) 72, 178, 181; American 172
Manihot esculenta 33
Mappia nimmoniana (*M. foetida*) 209
marigold, pot 191
marijuana 83, 153
masoprocol (mesonordihydroguaiaretic acid) 201, 214–15
maté 66
mayapple 161, 172–3, 200, 203–4; Himalayan 172–3, 200, 202–3; Indian 173, 202
mayflower 42
maypop 18
maytansine 22, 201, 212
Maytenus rothiana see *Gymnosporia rothiana*
Maytenus truncata 212
mazambron marron 165

meadowsweet 49, 138, 140, 142–3, 171, 173
Melaleuca alternifolia 128, 161, 170–1, 215
Melanthiaceae see trillium family
melilot 37, 48–9
Melilotus spp. 31, 37; *M. officinalis* 37, 48–9
Melissa officinalis 56, 59, 75
melon, bitter 114
Menispermaceae see moonseed family
menopause 183–6, 192–5, 197
Mentha spp., *M. aquatica*, *M. spicata*, 90; *M.* × *piperita*, 75, 81, 90–1, 138
menthol 90–1, 138, 145
menthone 90
Merrilliodendron megacarpum 209
metformin 103, 112–13
methadone 141
methoxsalen (8-methoxypsoralen) 168
methyl salicylate 143
methysticin 56
Metrosideros excelsa 128
mezereum family 33, 133
Middle Eastern medicine, introduction 13, 84–5
midwives' herb 153
miglitol 103, 115
migraine 58, 67, 191
milk bush 30, 47
milk/holy thistle 101–2, 104–5
milkweed 214
mind-altering 56–8, 69, 80, 83, 98, 108, 128, 140, 153, 156
mint 90–1; water 90
mint family 19, 32, 43, 58, 74–5, 81, 90, 97, 138, 145, 177, 185, 192, 205
Momordica charantia 103, 114–15
Monarda fistulosa 205
monoterpenes 19, 32, 90, 145, 170
moonseed family 17, 139, 152–3
Moraceae see mulberry family
morphine 20–2, 47, 77, 81, 96–7, 140–1, 155
Morus alba 114
moschatel family 127
motherwort 43
movement disorders 22, 29, 53–4, 70–1, 73, 113, 137–9, 152–4, 156, 181
Mucuna pruriens 154–5
mulberry, white 114
mulberry family 84, 115
Musaceae see banana family
mushrooms see fungi
Myristicaceae see nutmeg family
Myristica fragrans 15, 174
Myroxylon balsamum (*M. pereirae*) 160–1, 176
myrrh 84
myrtle family (Myrtaceae) 76, 118, 128, 161, 170, 174, 215

nabilone 83
nag kuda 209
naked ladies 148
naked-ladies family see autumn crocus family
naloxone 141
naphthodianthrones 60–1
Narcissus spp. 70–1
narcotine 125
neonicotinoids 69
neostigmine 73
nervous system 15, 18–19, 21–3, 31–3, 39–43, 47, 53–61, 64–77, 86, 89, 96,

102, 106, 108, 110, 120–1, 129, 137–9, 141, 147, 152–6, 161, 165–7, 169, 173, 175, 179, 181, 194, 196, 211
nettle, stinging 111
nettle family 111
New Zealand Christmas tree 128
nexehuac 151
Nicotiana spp. 33; *N. rustica* 151; *N. tabacum* 68–9, 151
nicotine 33, 68–9, 151
nightshade: deadly see belladonna; purple African 201
nightshade family see potato family
nigranoic acid 106
norethisterone (norethindrone) 186–7
North American medicine 204–5
Nothapodytes nimmoniana see *Mappia nimmoniana*
nutmeg 15, 174–5
nutmeg family 174
nux vomica 153
Nymphaea nouchalii var. *caerulea* (*N. caerulea*) 85
Nymphaeaceae see waterlily family
Nyssaceae see tupelo family

oat 164–5; wild 164
Oceanian medicine, introduction 128–9
Oenothera biennis 185, 193
Oleaceae see olive family
oleander, climbing 39
oleic acid 110
oleoresin 82, 145; aspidium 99
olibanum 85
olive family 62
omacetaxine 22, 207
Onagraceae see evening primrose family
onion 147; sea 13, 28
Ophiorrhiza pectinata 209
opiates, opioids 22, 92, 94, 96–7, 141, 155
opium 20–1, 47, 140–1
opium poppy 3, 13, 20–2, 36, 46–7, 77, 81, 96, 125, 137–41, 155
orchid family (Orchidaceae) 151
oseltamivir 119, 130–1
ouabain 39
ox tongue, prickly 80
oxidative stress (antioxidant) 19, 65, 75, 165, 216

paclitaxel 201–2, 206–7
Paeonia lactiflora 62
Paeoniaceae see peony family
pain (analgesics) 12, 20–2, 29, 47, 49, 56, 67, 88, 92, 96, 109, 138, 140–7, 164, 191
pain 15, 21, 29, 32, 43, 47, 49, 51, 55, 57–9, 61, 66–7, 70, 74, 76, 77, 82, 86–7, 90–1, 97, 105, 108, 114, 122–3, 127–8, 130, 137–8, 140–9, 151, 155, 165, 175, 181, 191, 195, 204–5, 211
palm: areca 98; date 85; sabal 110
palm family 80–1, 85, 103, 110, 129
Panax spp. 62, 65; *P. ginseng* 15, 54, 64, 106; *P. quinquefolius* 65
papain 167
Papaver somniferum 3, 13, 21, 36, 77, 81, 125, 137–8, 140–1
Papaveraceae see poppy family

papaverine 47, 125
papaya 167
papaya family 167
paperbark, narrow-leaf 170
papyrus sedge 13
parasites 13, 19, 32, 41, 45, 47, 79–80, 98–9, 113, 117, 119, 128, 134–5, 159–61, 173, 176–7, 188
pareira brava 152
Parkinson's disease 22–3, 31, 47, 57, 72, 139, 154–6, 181
parsley 29, 143, 191
parsnip (*Pastinaca sativa*) 29
passion flower (*Passiflora incarnata*) 18
passionfruit family (Passifloraceae) 18
Paullinia cupana 66–7
Pausinystalia johimbe see *Corynanthe johimbe*
pea family see legume family
Papyrus, Ebers 13, 86, 200
Pedaliaceae see sesame family
peganine 119
peltatins 173
pennywort: Asiatic 166–7; Indian 167
peony 62
peony family 62
pepper family 45, 54, 56, 85, 99, 129
peppermint 75, 81, 90–1, 138, 145
peppers 128, 144–5; black 45, 85; chilli 33, 138–40, 144, 151; intoxicating/narcotic 57; monk's 192–3
periwinkle, lesser 75; Madagascar 22, 199, 201–2, 209–11
periwinkle family see dogbane family
Petroselinum crispum 29, 143, 191
phenformin 113
phloroglucinols 60–1, 99
Phoenix dactylifera 85
pholcodine 141
Phormium tenax 128
Physostigma venenosum 21, 55, 72, 108, 161, 179
physostigmine 21, 55, 72–3, 178–9
phytoestrogens 195
phytosterol 110
pilocarpine 98, 178–9
Pilocarpus jaborandi 98, 179; *P. microphyllus* 161, 178
Pimelea prostrata 133
Pimpinella anisum 29
pine (*Pinus* spp.) 97
pine family (Pinaceae) 97
pineapple 167
Piper betle 99, 129; *P. excelsum* see *Macropiper excelsum*; *P. methysticum* see *M. methysticum*; *P. nigrum* 45, 85
Piperaceae see pepper family
piperine 45
Plantaginaceae see speedwell family
Plantago spp., *P. ovata* 81
plantain 81
plantain family see speedwell family
Plectranthus barbatus see *Coleus barbatus*
Pliny the Elder 105, 174, 206
Poaceae see grass family
podophyllotoxin, podophyllin resin 172–3, 200, 202–4
Podophyllum hexandrum 172–3, 200, 202; *P. peltatum* 161, 172, 200, 203–4
pōhutukawa 128

poison: bulb 109; Somali arrow 37, 39, 108–9
poisons/toxins 8, 12, 17, 19, 24, 29, 30–1, 33, 37, 39–40, 46–8, 55, 57, 59, 68–9, 71–3, 77, 80–2, 87, 91, 94, 97, 102, 104–5, 108–9, 112–13, 115, 122, 127, 131, 133, 135, 138, 141, 148–9, 152–3, 155–6, 172–3, 177, 179–81, 185, 189, 200, 203, 206, 208, 212, 214–15
polyacetylenes 29
Polygonaceae see knotweed family
polysaccharides 19, 31, 164–5
pomegranate 84
pomegranate family 84
poplars (Populus spp.) 12
poppy family 13, 36, 77, 81, 125, 138, 140
potato 33, 128, 150, 173
potato family 13, 21, 33, 45, 65, 68, 72, 80, 82, 86, 91, 138–9, 144, 150, 154, 156, 161, 178, 181, 201, 215–16
proanthocyanidins, procyanidins 33, 42, 94
procaine 76–7
progesterone 186–7, 194
prostate, benign prostatic hyperplasia 101–3, 110–11, 195, 216
prostratin 119, 132–3
protea family (Proteaceae) 129
protoveratrines 36, 46
Prunus africana 110–11
Psoralea corylifolia see Cullen corylifolium
psoralens 29, 160, 168–9
psoriasis 31, 160, 164, 168–9, 171, 179
pukeweed 69
Punica granatum 84
Putterlickia pyracantha 201; P. verrucosa 212
pygeum 111
Pygeum africanum see Prunus africana
pyranocoumarins 124
pyrethrin 55

qing hao 134–5
quercetin 42
quinidine 40–1
quinine 20, 41, 134–5

Ranunculaceae see buttercup family
raspberries, black 217
Rauvolfia serpentina 36, 40, 61; R. tetraphylla 47; R. vomitoria 40
redbush 108
reproductive system 28, 32, 40, 43, 65–6, 84, 87, 102–3, 105, 110, 112–13, 115, 143, 155, 163, 172, 183–9, 192–7, 212
reserpine 36, 40, 47, 61
resins 84, 95–7, 122–3, 172–3, 175, 177, 202, 204–5, 215
respiratory conditions 18–19, 31, 41, 47, 51, 57, 62, 66–7, 69, 74, 82, 86, 89–91, 96, 105, 108, 117–27, 130–1, 140–1, 143–5, 148–9, 156, 168–70, 175, 177, 179, 205, 212, 214–15
Rhamnaceae see buckthorn family
Rhamnus cathartica 93; R. purshiana see Frangula purshiana
rhein anthrone 93
rhubarb (Rheum palmatum) 18
rice flour, prostrate 133

ricinoleic acid 33
Ricinus communis 12–13, 177
rituals see ceremonies
rivastigmine 21, 55, 70, 72–3
rooibos 108
rose family (Rosaceae) 37, 42, 49, 111, 138, 217
roselle 18–19
rosemary 75, 145
Rosmarinus officinalis see Salvia rosmarinus
rotundifuran 192
rubber tree 95
Rubiaceae see coffee family
rubitecan 209
Rubus occidentalis 217
Rudbeckia spp. 127
Rumex crispus 168–9
Ruscus aculeatus 28, 37, 51
Rutaceae see citrus family
rye 155

Saccharum officinarum 129
safflower 115
saffron 60–1, 115; meadow 148
sage 75; Chinese 37, 43
Salicaceae see willow family
salicin 142–3
salicylates 12, 21, 48–9, 142–3, 171, 173
salicylic acid 142–3, 171, 173, 215
Salix spp. 12, 142; S. alba 21, 37, 138, 142, 161
Salvia miltiorrhiza 37, 43; S. officinalis 75; S. rosmarinus 75, 145
Sambucus nigra 126–7
sangre de drago/grado 94
santonin 99
Sapindaceae see soapberry family
sapogenin (aglycone) 184, 186; steroidal 186
saponins 28, 64; steroidal 28, 51, 65, 114, 186
sarpagandha 40
sarsaparilla 163; Mexican 163, 186
sarsasapogenin 163, 186
saw palmetto 103, 110–11
Schisandra spp. 62, 107; S. chinensis 102, 106–7; S. glabra 107
Schisandraceae see magnolia vine family
schisandrins 106
scopolamine see hyoscine
Scopolia spp. 91; S. carniolica 80, 86; S. japonica 86
scyllo-cyclohexanehexol 22
Secale cereale 155
sedatives 56, 58–9, 86–7, 109–10, 140–1
sedge family 13
seirogan 94, 96
senna 13, 18, 31, 92–3; Alexandrian/Indian/Tinnevelly 92–3
Senna spp. 92; S. alexandrina 13, 92–3
sennosides 92–3
Serenoa repens (S. serrulata) 103, 110–11
Seriphidium cinum 99
serpentwood, African 40
sesame family 109, 138, 146
sesquiterpenes 58, 74–5, 122, 131, 134–5, 167, 188
shatavari 45
Shen Nong Ben Cao Jing (The Drug Treatise of the Divine Countryman) 16, 82, 107, 120
shikimi 131

shikimic acid 119, 130–1
shogaols 82
Shorea spp. 97
sickness, nausea, vomiting 76, 79–80, 82–3, 86–7, 91, 108
Siddha medicine, introduction 44–5, 106
silibinin (silybin) 102, 104–5
Silybum marianum 101–2, 104–5
simplexin 133
sinecatechins 25, 172–3
Sinopodophyllum hexandrum see Podophyllum hexandrum
sisal 28, 125, 160, 162–3
sitosterol, beta- 110–11
skin 12, 22–3, 29, 31–3, 38–40, 45, 48, 51, 58, 60, 62, 77, 87, 94–5, 109, 113–14, 126–9, 138, 142–5, 159–73, 175–9, 181, 191, 193, 201–3, 205, 214–15
Smilacaceae see catbrier family
smilagenin 22, 163
Smilax spp. 163; S. aristolochiifolia 163, 186
smoking 68–9
snakeroot: black 194–5; Indian 36, 38, 40, 47, 61
snowdrops 55, 70–1; common 3, 70; Woronow's 70
snowflakes 70; summer 71
soapberry family 37, 50, 66
sodium cromoglicate 118, 124–5
solamargine 201, 215
Solanaceae see potato family
Solanum spp. 33; S. lycopersicum 33, 151, 216; S. marginatum 201; S. melongena 33; S. tuberosum 33, 150
solasodine glycosides 201, 215
solasonine 215
soursop family 128, 175
South American medicine, introduction 150–1
sparteine 40
spearmint 90
speedwell family 21, 36, 38, 81, 167, 205
spiderwort family 103
spike-thorn: false 201; mock forest 212; Roth's 212
spilanthol 77
spindle family 22, 201, 212
spurge, petty 22, 33, 201, 214–15
spurge family 13, 22, 33, 81, 94–5, 119, 132–3, 177, 201, 214
squill, sea 13
St John's wort 11, 18, 54, 60–1
St John's wort family 18, 54, 60
star anise 56, 130–1; Chinese 130; Japanese 131
star vine 107
starflower 185
Stephania tetrandra 17
steroids 28, 160, 162–3
stimulants 31, 55, 62, 64, 66–7, 69, 77, 98, 108, 118
storax family 175
stramonium 156
Strathmore weed 133
strophanthidin 39
Strophanthus gratus 39; S. kombe 39, 109
strychnine 153
Strychnos nux-vomica, S. toxifera 153
Styracaceae see storax family
Styrax benzoin 175
sugar cane 129
sulforaphane 147
surgery 55–6, 58, 71, 77, 110, 138–9, 152–3, 181

swamp cypress family 97
Syzygium aromaticum 76, 174

Tabernaemontana alternifolia 209; T. iboga 69, 108
Tamiflu 131
Tanacetum coccineum 55; T. parthenium 191
Taxaceae see yew family
taxane 207
Taxodiaceae see swamp cypress family
taxol 206
Taxus spp. 121, 206; T. baccata 206–7; T. brevifolia 201, 206–7
tea 25, 55, 62, 66; tree 128–9, 161, 170–1, 176–7, 215; desert 120, 205; green 95, 147, 172–3; Mormon 118, 205; Paraguay 66; teamster's 205
tea family 55, 62, 147, 172
teak 97
Tectona spp. 97
temperature/fever 21, 41, 49, 51, 54, 57, 66, 86, 91, 113–14, 118, 124, 130, 132, 134–5, 142–3, 159, 179, 191
teniposide 202–3
tequila 163
Terminalia bellirica 45, 85
terpinen-4-ol 170–1
testosterone 110, 186–7
tetrabenazine 156
tetrahydrocannabinol, delta-9- (THC) 80, 83, 153
Theaceae see tea family
thebaine 141, 155
Theobroma cacao 31, 151.
theobromine 31
theophylline 31, 66–7, 125
thorn apple 139, 154, 156–7
thyme 19, 177
thymol 19, 205
Thymelaeaceae see mezereum family
Thymus vulgaris 19, 177
tigliianes 33
Tilia cordata 31
tlapalcacauatl 151
tlilxochitl 151
tobacco 33, 68–9, 89, 99, 111, 151; Aztec/coarse 151; Indian 69; sweet 151
tokoro 186
tolohua xihuitl 151
tomato 33, 151, 216
toothache plant 77
toothpick weed 124–5
topotecan 21, 201–2, 208–9
toxiferine 153
trastuzumab-DM1 (trastuzumab emtansine) 22, 212
Trifolium pratense 195
Trigonella foenum-graecum 187
Trillium erectum 195
trillium family 36, 46, 195
triterpenes and their glycosides 22, 31, 50–1, 88, 106, 114, 146, 166, 194
tubocurarine 139, 152–3
tupelo family 21, 201, 208
turmeric 45, 102, 104, 106–7, 147, 201, 216–17

ulcers: gastrointestinal 79–80, 88–9, 95, 164; leg 39, 113, 177
Umbelliferae see carrot family
Unani medicine, introduction 44–5
urinary system 17–18, 28, 40, 64, 66, 105, 114, 124, 153, 169, 191, 195

Urtica dioica 111
Urticaceae see nettle family
uzara 30

Vaccinium spp. 217
valepotriates, valerenic acid 58
valerian (Valeriana officinalis) 54, 56, 58–9
Valerianaceae see honeysuckle family
vanilla (Vanilla planifolia) 151
varenicline 69
vasaka 119
vasicine see peganine
Vataireopsis araroba 31, 160, 169
velvet leaf 153
verapamil 36, 47
Veratrum album 36, 46; V. viride 46–7
Veronicastrum virginicum 205
verrucas 172–3
Viburnaceae see moschatel family
Vicia faba 154
vinblastine, vincristine, vindoline 22–3, 201–2, 210–11
Vinca minor 75
vincamine 75
vindesine, vinflunine, vinorelbine 210
vinpocetine 75
viruses 65, 70–1, 94, 105, 117–19, 126–7, 130–3, 145, 160, 171–3, 188
visnadin, visnagin 124
Visnaga daucoides 36, 118, 124–5
Vitex agnus-castus 183, 185, 192–3
vitexin 42
vitiligo 45, 168
volatile oils see essential (volatile) oils

warfarin 37, 48–9
wart grass/wort 214
warts 62, 172–3, 202–3, 214
waterlily family 85
white pear family 209
willow 12–13, 21, 23, 37, 48–9, 138, 140, 142–3, 161, 171, 173; white 142
willow family 12, 21, 37, 138, 142, 161
wintergreen 143
witch hazel 62, 204–5
witch hazel family 62, 205
Withania somnifera 33, 45, 65
withanolides 65
wormseed, Levant 99
wormwood: Levant 99; sea 99; sweet 23, 63, 119, 134–5
wu wei zi 107

Xysmalobium undulatum 30

yam 125, 163, 184, 186–7; Mexican 185–7; wild 186–7
yam family 125, 163, 184, 186
yangonin 56
yew 121; English/European 206–7; Japanese plum 22, 207; Pacific 201–2, 206–8
yew family 22, 201, 206
yin yang huo 197
ylang ylang 128, 175
yohimbe 108, 185, 196–7
yohimbine (corynine, quebrachine) 40, 196–7
Yucca spp. 22, 163

Zingiber officinale 45, 80, 82–3, 106
Zingiberaceae see ginger family
Zygophyllaceae see caltrop family

Acknowledgements

The authors would like to thank colleagues at the Royal Botanic Gardens, Kew for checking, within their respective fields, the plant names and images included in this book, including Dr D. J. Nicholas Hind (Asteraceae), Dr Gwilym Lewis (Fabaceae), Dr Alan Paton (Lamiaceae), Dr David Goyder (Apocynaceae), Dr Christine Leon MBE (traditional Chinese medicines), Dr Rafael Govaerts (plant nomenclature) and Dr Aljos Farjon (Gymnosperms). The authors would also like to thank Dr Peter Gasson, Dr Gwilym Lewis, Dr Martin Cheek, Dr D. J. Nicholas Hind and Dr Aljos Farjon for permission to use their wonderful and unique photos in this book. They acknowledge the many scientists whose work has been consulted and drawn on.

The authors also thank the fantastic editorial team at Bright Press, Jacqui, Judith, David and Susi, and designer Tony, for their support throughout the process of writing and producing this book, and colleagues in Kew Publishing, especially Gina Fullerlove, Lydia White and Pei Chu, for their support and publishing expertise, which has been much appreciated during the development of this book.

Liz could not have written this book without the support and encouragement of her parents, husband and daughter.

Melanie-Jayne would like to thank her family, especially her parents, for their support throughout writing this book; and Emeritus Professor Peter J. Houghton of King's College London, who has been an inspirational pharmacognosist throughout her career.

Picture credits

All images copyright the following (T = top, B = bottom, L = left, R = right):

Alamy Stock Photo: 2 blickwinkel; 5 David Gowans; 8 Zoonar GmbH; 13 Alon Meir; 18 Gerry Bishop; 24 Zoonar GmbH; 30T yogesh more; 33L Maximilian Weinzierl; 37B Xinhua; 41B Science Photo Library; 44 Science History Images; 47 Janet Horton; 48 Frank Hecker; 51 Bob Gibbons; 63B Organica; 72B The Natural History Museum; 77 Robert Biedermann; 84 Historic Images; 9 & 87B AY Images; 92 WILDLIFE GmbH; 108 Eric Nathan; 124 Frank Hecker; 125T Goran Bogicevic; 129T jackie ellis; 134 Frank Hecker; 143B WILDLIFE GmbH; 147 (main image) Arterra Picture Library; 150 Island Images; 151T Bob Gibbons; 155T Manfred Ruckszio; 165T marilyn barbone; 168 Tim Gainey; 175B Hemis; 180R Scott Camazine; 185B Hemis; 190 Arco Images GmbH; 195B 503 Collection; 202 Anna Yu; 204B Manfred Ruckszio; 209L John Lander; 214 Premaphotos; 215 Robert Shantz; 217 Vainillaychile.
The Board of Trustees of the Royal Botanic Gardens, Kew: 26T, 26B, 27.
Dr Martin Cheek: 67B.
Dr Elizabeth A. Dauncey: 12, 172L.
Dreamstime: 45T Seksan Panpinyo; 71B Gmlynch; 95R Madeleinesteinbach; 196 Kanusommer.
Dr Aljos Farjon: 22.
Dr Peter Gasson: 42B, 179, 180L, 207.
Dr Tom Gregory: 66T, 98.
Manuel de Jesús Hernández Ancheita: 186.
Dr D. J. Nicholas Hind: 81L.
Dr Gwilym Lewis: 169B.
Andrew McRobb/RBG Kew: 210.
Dr Craig Peter: 211.
Science Museum, London. CC BY: 21B, 135B.
Science Photo Library: 16B Andy Crump; 20L Leonard Lessin; 21T National Library of Medicine; 72T Jerry Mason; 76L Michael W. Davidson; 94 Dr Morley Read; 95L Dr Morley Read; 132R Steve Gschmeissner; 153 Ted Kinsman/Science Source; 161T Dr Fred Hossler, Visuals Unlimited; 177R

Dennis Kunkel Microscopy; 189B Martin Shields; 200 Alfred Pasieka; 204T Oxford Science Archive/Heritage Images; 206 Alan Sirulnikoff.
Shutterstock: 14 Nila Newsom; 15T kokixx; 15B Nataly Studio; 17B chengyu; 19 Passakorn Umpornmaha; 23T Zocchi Roberto; 23B Heavy Rain; 25 Lotus Images; 28L Simic Vojislav; 28R msgrafixx; 29L Starover Sibiriak; 29R Scisetti Alfio; 31L Mukund Kumar; 31R Nick Pecker; 33R tinglee1631; 37T alybaba; 38 Alexander Varbenov; 42T bogdan ionescu; 43T domnitsky; 43B LianeM; 45B Valentyn Volkov; 46 Catalin Petolea; 49T Marian Weyo; 49B dabjola; 50 Whiteaster; 54L whitehoune; 54R august0809; 55L Tatyana Mi; 55R Mirabelle Pictures; 56 goatcafe; 57B ChameleonsEye; 59TL Vaclav Mach; 59TR Richard Peterson; 60 Ines Behrens-Kunkel; 61T ThomasLENNE; 62 Nickolai Repnitskii; 63T Dragon Images; 64 Igor Cheri; 65T Indian Food Images; 66B Doikanoy; 67 guentermanaus; 68 LianeM; 70 Elena Koromyslova; 71T yakonstant; 74 Slobodan Kunevski; 75T Anton Zhuk; 75B Pixeljoy; 80 Volodymyr Nikitenko; 81BR UbjsP; 82 jiangdi; 83B Opra; 85T WeihrauchWelt; 9 & 86 Miroslav Hlavko; 89T Antonio Gravante; 89BL Manfred Ruckszio; 90T Dionisvera; 90B Burkhard Trautsch; 93T picturepartners; 93B Robert Mutch; 96 tk312001; 97T Anitham Raju Yaragorla; 97B jaroslava V; 99T pisitpong2017; 99B Keith Michael Taylor; 102 NNphotos; 103 Charles G. Haacker; 104 jaroslava V; 105T el_cigarrito; 106 alexmak7; 107T Lotus Images; 107B JIANG HONGYAN; 110 Jason Patrick Ross; 111T BW Folsom; 114 Swapan Photography; 115T Muellek Josef; 115B Phichai; 119 Debasis Das; 120 Yana Kho; 121 NOPPHARAT STUDIO 969; 122 Everett Historical; 123 k_lmnop; 126 JRP Studio; 127B spline_x; 128 patjo; 129B Molly NZ; 130 leungchopan; 131 Kateryna Kon; 139B TMsara; 140 Mr.Jinda Lungloan; 141T Daniel Prudek; 141B Marzolino; 143T dabjola; 144L joysugarplum; 146 Ademoeller; 148 TonelloPhotography; 149 Wiro.Klyngz; 155B wasanajai; 161B Scisetti Alfio; 162 (main image & inset) COULANGES; 163 Lamax; 164 DeawSS;

166L DESIGNFACTS; 166R Nataliia Melnychuk; 167 Piyachok Thawornmat; 169T simona pavan; 170 tamayura; 171T Ebrahim98; 171B VIRTEXIE; 172R Giel; 174 anilkumart; 175T johan kusuma; 176 Skyprayer2005; 177L Emilio100; 181T Heike Rau; 181B Watercolor_swallow; 185T vainillaychile; 187B Swapan Photography; 188 Sai Nam; 191B Nadezhda Nesterova; 192 simona pavan; 193T shansh23; 193B unpict; 194 FAG; 195T Melinda Fawver; 197 JIANG TIANMU; 201B Mike Russell; 205L LianeM.
Dr Wolfgang Stuppy/RBG Kew: 36.
Wellcome Collection. CC BY: 16T, 59B, 65B, 113B, 151B, 152, 187T.
Wikimedia Commons. CC 2.0: 61B Dinesh Valke; 118 Andrey Zharkikh; 125B Dinesh Valke; 145 THOR; 184T Vahe Martirosyan; 203 Nicholas A. Tonelli.
Wikimedia Commons. CC 2.5: 105B Valérie75; 111B Marco Schmidt.
Wikimedia Commons. CC 3.0: 73B Keisotyo; 81TR Stan Shebs; 9 & 87T Luca Fornasari; 89BR Michael Wolf; 127T H. Zell; 142 (main image) Willow; 142 (inset) S Sepp; 147 (inset) Roger Culos; 165B Jnoel974; 201T Michael Wolf; 205R H. Zell; 208 Geographer; 209R Vinayaraj.
Wikimedia Commons. CC 4.0: 30B SAplants; 32L Krzysztof Ziarnek; 39 Vengolis; 41T Andreas Eichler; 85B Vengolis; 109B Italian Boy; 112 Stefan.lefnaer; 133T Mark Marathon; 133B Krzysztof Ziarnek; 157 Stefan.lefnaer; 184B Krzysztof Ziarnek; 213T Vinayaraj; 213B Vinayaraj.
Lin Yu-Lin: 17T.

All other images in this book are in the public domain.

Every effort has been made to credit the copyright holders of the images used in this book. We apologize for any unintentional omissions or errors, and will insert the appropriate acknowledgement to any companies or individuals in subsequent editions of the work.